KU-497-303

Clouds in a Glass of Beer

Simple Experiments in Atmospheric Physics

CRAIG F. BOHREN

Pennsylvania State University

Foreword by
JEARL WALKER

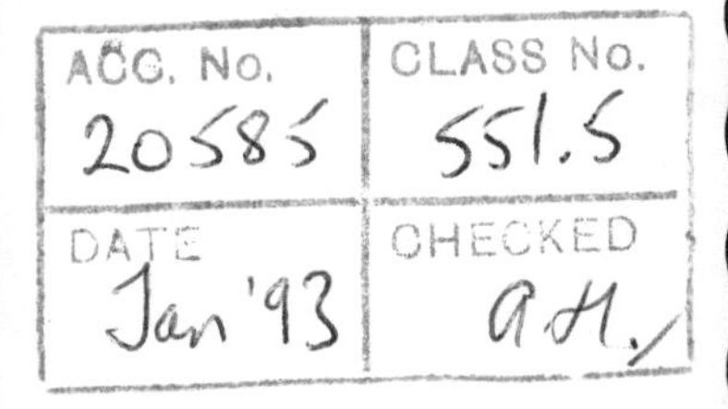
ACC. No. 20585
CLASS No. 551.5
DATE Jan '93
CHECKED a.H.

WILEY SCIENCE EDITIONS

JOHN WILEY & SONS, INC.

New York · Chichester · Brisbane · Toronto · Singapore

Permissions

Bohren, Craig, "Nonrandom Bubbles," *Science,* Vol. 216, #4547, p. 682 (Letters section), 14 May 1982. Copyright © 1982 by the American Association for the Advancement of Science.

Weather, Vol. 15, p. 245 (1960). Royal Meteorological Society, James Glaisher House, Grenville Place, Bracknell, Berkshire RG12 1BX.

DENNIS THE MENACE ® used by permission of Hank Ketcham and © by North America Syndicate.

Laurens van der Post, "Venture to the Interior," William Morrow & Co., 105 Madison Ave, NY, NY 10016.

© Sydney Smith, *Mostly Murder,* first published in Great Britain 1959 by Harrap Ltd., 19-23 Ludgate Hill London EC4M 7PD.

Publisher: Stephen Kippur
Editor: David Sobel
Managing Editor: Andrew Hoffer
Production Service: G&H SOHO, Ltd.

Copyright © 1987 by John Wiley & Sons, Inc.

All rights reserved. Published simultaneously in Canada.

Reproduction or translation of any part of this work beyond that permitted by section 107 or 108 of the 1976 United States Copyright Act without the permission of the copyright owner is unlawful. Requests for permission or further information should be addressed to the Permission Department, John Wiley & Sons, Inc.

Library of Congress Cataloging-in-Publication Data

Bohren, Craig F., 1940-
Clouds in a glass of beer.

(Wiley science editions)
1. Atmospheric physics. 2. Atmospheric physics—Experiments. I. Title. II. Series.
QC861.2.B64 1987 551.5 87-13375
ISBN 0-471-62482-9

Printed in the United States of America

88 10 9 8 7 6 5

Clouds in a Glass of Beer

Clouds in a Glass of Beer
: Simple Experiments in

The Wiley Science Editions

The Search for Extraterrestrial Intelligence, by Thomas R. McDonough
Seven Ideas That Shook the Universe, by Bryon D. Anderson and Nathan Spielberg
The Naturalist's Year, by Scott Camazine
The Urban Naturalist, by Steven D. Garber
Space: The Next Twenty-Five Years, by Thomas R. McDonough
The Body in Time, by Kenneth Jon Rose
The Complete Book of Holograms, by Joseph Kasper and Steven Feller
The Scientific Companion, by Cesare Emiliani
Starsailing, by Louis Friedman

Dedicated to the memory of Louis J. Battan.
This book would not have been written
had our paths never crossed.

Foreword

Suppose you had the opportunity of peering over the shoulder of Sherlock Holmes during an investigation. Imagine him examining what appears to be an ordinary scene. You and he see the same things, yet only he spots the clue that unravels the mystery.

Bohren is like Holmes in that he perceives mystery and puzzlement in what most of us would think is a common, uninteresting setting. You and he see the same things, but he notes the curious feature that unlocks some marvelous demonstration of scientific principles. This book gives you the opportunity of peering over Bohren's shoulder as he explores the world. He will take you from tavern to countryside and from a greenhouse to a murder scene, all the while teasing you into recognizing the curious clues along the way.

The experience will be rewarding and lasting. From now on, every time you see a rainbow, notice the darkening of wet sand, discover your bathroom mirror covered with drops, or note the coloring of the sky during sunset, you will recall this book. Once exposed to Bohren's investigative skills, you will be unable to take the world for granted. The steam rising from a bowl of hot soup, the freezing of a country pond, and the bobbing of a toy duck will return you to Bohren's teachings. And every time you overhear someone rattling off, "Well, that happens only once in a blue moon," you will remember how the moon might indeed turn blue.

Bohren has a second purpose here. He greatly enjoys tilting with the legends and blatant falsehoods that have crept into our casual descriptions of the physical world. Does warm air really *hold* more water vapor than cool air? Does a greenhouse actually grow warm because it traps infrared radiation? Does frost fail to develop on the grass beneath a tree because the tree shields the grass? Does sunlight actually heat the air around us? These and other questions have been answered wrongly countless times. Bohren sets us straight on the answers, taking care that we understand the principles sufficiently so that we never repeat the mistakes.

The content of the book runs the spectrum from serious to playful. However, throughout, it rings with a unifying tone: the science of the everyday physical world is fun. And so is this book.

Jearl Walker

Physics Department,
Cleveland State University

author of The Flying Circus of Physics
and "The Amateur Scientist," monthly in Scientific American

Preface

It is strange that I find myself writing this book. When I was younger I fancied myself as a theoretical physicist, that is, a manipulator of cabalistic symbols. Toward the middle of 1978, however, I found myself in somewhat of a bind. I was about to end three years as an expatriate in Wales and return to the United States, where my prospects seemed bleak.

But on a brief visit to the University of Arizona I happened to run into Louis Battan, the late director of the Institute of Atmospheric Physics. When in the course of our conversation he learned that one of the purposes of my visit was to find work, he asked me if I would like to teach elementary meteorology in the fall. I said that I didn't know anything about meteorology. "That's all right," he responded without flinching, "neither do the students." And so it was that in September I found myself hearing for the first time about meteorology—from my own mouth.

My students were "nonscience" majors, although "anti-science" would perhaps be closer to the mark. It didn't take me long to realize that I could not reach them by the kinds of arguments and approaches that appealed to me. The merest whiff of an equation would stampede them to the dean's office to complain about cruel and unusual punishment. Carefully crafted syllogisms moved them about as much as it would have moved a team of sled dogs. What did move them, I discovered, were things they could see and hear and touch. So I began trying to devise simple demonstrations to help convey physical ideas. I made a point of always bringing some object—a glass of water, an iron bar—to class with which to punctuate my lectures. The success of each demonstration stimulated me to devise more.

When I moved to Pennsylvania in 1980 to take a job teaching meteorology at the Pennsylvania State University, I took with me my newly acquired habit of incorporating demonstrations into my lectures. With time I came to realize that these demonstrations were as much for my delight as for the students'. My tastes were shifting from the abstract to the concrete. As so often happens to those who have converted, especially late in life, I embraced demonstrations with single-minded passion. Of necessity, I made use of things I could fish out of wastebaskets or find in my home. My budget for equipment was zero. A high-school physics teacher has more equipment than I dare dream of. A visitor once asked to see my "laboratory." I showed it to him without comment: a tiny darkroom, which I have to defend continually against those who would use it as a storage closet.

Necessity aside, there is a virtue in simple demonstrations and experiments, the kind that everyone can do for themselves without special or expensive equipment. Some science makes use of such complicated equipment that it is intimidating. To do it, or even to understand it, is like sinking a deep mine shaft in hard rock using enormous and expensive machinery. Yet there are many scientific nuggets almost on the surface, accessible to all who would merely scratch for them. It is this kind of science that is the subject matter of this book. Nowhere in it do I invoke the revelations of authorities to convince you of anything. In science there is no authority other than observation and experiment illuminated by reason. I urge readers to do their own experiments and make their own observations. I neither invoke authority nor would I wish to be invoked as one. Authorities do have legitimate uses, but not as weapons. The pronouncements of authorities should be weighed carefully—and then rejected if found wanting.

I got into meteorology by accident. And this book owes its existence to yet more accidents. During my first year at Penn State, a freshman meteorology student, Gail Brown, approached me about taking an elementary course I was teaching. It was too elementary for her, so to make it more challenging I asked her to do everything the other students had to do and in addition to write one-page summaries of my demonstrations. This she did, and presented them to me bound together at the end of the semester. Not long after, Linda Dove, then the managing editor of *Weatherwise*, was looking for someone to write a regular column on experiments in atmospheric physics. We sent her Gail's compendium, and as a result we became columnists. When Gail graduated, I continued alone.

This book is distilled from the articles I published, either by myself or with Gail as co-author, between 1981 and 1987. I have completely rewritten all of them in light of my own second thoughts and comments from readers and have organized them into a coherent whole.

The subtitle of this book is Simple Experiments in Atmospheric Physics. Atmospheric physics, loosely defined, is all of atmospheric science not directed toward understanding atmospheric motions, especially those motions (winds) that bring changes in the weather. A distinction is usually made between atmospheric physics and meteorology, although like most such distinctions it is artificial: nature surely does not recognize it. Some of the members of my department are directly concerned with weather forecasting. They do what most people think that meteorologists do. I, on the other hand, couldn't forecast the weather if my life depended on it. I am a physicist who happens to have found a home in a meteorology department, so if a label had to be attached to me it would be "atmospheric physicist."

The explanations in the following chapters are thorough. I do not shrink from discussing subtleties. I have labored long and hard to make my explanations as simple as possible, by which I mean nonmathematical and not strangled in jargon, but not at the expense of making them simplistic. My aim was simple explanations, but not ones that would, in the words of James Clerk Maxwell, later require my readers "to unlearn a science that they might at length begin to learn it."

Acknowledgments

Gail Brown and I began writing the articles on which this book is based in 1981 at the suggestion of Alfred Blackadar, who introduced us to Linda Dove, then the managing editor of *Weatherwise*. Editors can influence what writers produce, for better or for worse. I am pleased to say that Linda did everything in her power to make writing articles as pleasant as possible. She has the rare ability to get the best out of people. She heroically resisted the temptation to meddle. And she never censored anything I wrote. At most, she occasionally counseled me to soften some of my harsher sentences. Her successor, Patrick Hughes, has continued in her tradition.

At Penn State I have relied upon many of my colleagues. I am sure they will understand that although I am grateful to all of them, Alistair Fraser's help has been paramount. I incorporate his ideas and his marvelous photographs in my manuscripts, and then audaciously ask him to review the results. He has critically reviewed almost every manuscript I have written, purging them of errors, clarifying muddled ideas, and making valuable suggestions. He, more than anyone else, has been the anvil on which I have forged my thoughts.

Everything I write is submitted to colleagues for comments, the more critical the better. I would rather be pummeled in private than in public. I am blessed with colleagues who are knowledgeable, unsparing with their time, and severe critics. At Penn State I have from time to time relied on critical reviews by Dennis Thomson, Chris Fairall, Bruce Albrecht, Bill Frank, Rosa de Pena, John Olivero, and Tim Nevitt (now at 3M). I have also profited greatly from discussions about cloud physics with Jorge Pena. Although Bill Mach is not an official member of the Penn State faculty (he is at Florida State), his connections with this institution are sufficiently strong that I must include him in the list of my colleagues there.

I am indebted to another former Penn State student, Greg Stone (now at Los Alamos), whose example I followed in making the cooling curve measurements discussed in Chapter 8.

I also thank the employees of The Camera Shop in State College for their help. They have lent me equipment, given me advice, and rushed film processing so that I could meet deadlines.

Because I spend my summers away from State College, I have had to submit manuscripts to my wider circle of colleagues for review before sending them off to the publisher: at Dartmouth, Bill Doyle; at the Cold Regions Research and Engineering Laboratory, Sam Colbeck; at the Naval Ocean Systems Center, Brian Thomason and Ray Noonkester. To them, also, I am grateful.

During my summers at Los Alamos, made possible by Gary Salzman and the late Paul Mullaney, I have leaned on still another set of colleagues. Shermila Singham and Roger Johnston each contributed a figure to this book, and this is acknowledged where they appear.

I can't mention all the people who have written to me, although I must say that I have appreciated their letters. Yet I must thank at least one correspondent by name. Duncan Blanchard (SUNY, Albany) was encouraging from the outset. He wrote to us after our first articles were published, and he continues to write letters in which encouragement is blended with criticism. Duncan is a superb writer of popular science, so praise from him has meant much.

It will become apparent to anyone who reads the following chapters that I rely heavily on colleagues not only for criticism but for ideas as well. Wherever an idea is not my own, I try to give its source. If I have failed to do so, it is a fault of memory not of will. Several of my colleagues at the University of Arizona contributed to my articles, hence this book, often unwittingly. In particular, I am grateful to Sean Twomey, Donald Huffman, John Kessler, and Philip Krider. Although Louis Battan did not have a direct hand in the preparation of my articles, his influence can be felt in all of them, and it is to his memory that this book is dedicated.

I wrote this book during a sabbatical leave in the Department of Physics and Astronomy at Dartmouth College. I am grateful to John Kidder, John Walsh, and Bill Doyle for making this possible. My primary reason for coming to Dartmouth was to learn as much as I could from Bill. He has contributed substantially to this book, both as a reviewer of manuscripts and of parts of chapters. And he has inspired me to tighten my arguments and rid them of unnecessary trimmings.

Several chapters were critically reviewed in draft form: Chapter 1 by Duncan Blanchard; Chapter 4 by Milton Kerker (Clarkson University); Chapter 12 by John Kidder; Chapter 21 by Alistair Fraser. The best way to thank them is to say that I tried to follow their suggestions.

Other members of the community in and around Dartmouth helped in various ways. As you will see in Chapter 9, I made good use of conversations with George Ashton at the Cold Regions Research and Engineering Laboratory. Phillip Fischer and Ron Ghosh tried to absorb the shocks I received in my bouts with the computer. Ralph Gibson helped in so many ways that I have lost track of them. Carol Selikowitz spent hours in her darkroom transforming my film into prints.

At Wiley I thank Beatrice Shube for acting as a matchmaker between me and David Sobel, who oversaw the publication of this book. Both he and his assistant, Dawn Reitz, patiently answered dozens of questions and tried to relieve me of some of the chores I would have fumbled. The production stages of this book were in the able hands of Andrew Hoffer of Wiley and Claire McKean of G&H SOHO, Ltd. I am grateful to Claire for light-handed, yet constructive, editing of the original manuscript.

Heldref Publications granted me permission to use my articles, for which I thank them.

Finally, I express my heartfelt gratitude to my companion of 23 years, Nanette Malott Bohren, who has served as experimental subject, photographer's model and assistant, critic, and proofreader. She told me once that "you'll do the next book by yourself." But she didn't mean it.

Hanover, New Hampshire
April 1987

Contents

It may be also useful to have results of mathematical reasoning expanded into ordinary language for the benefit of mathematicians themselves, who are sometimes too apt to work out results without sufficient statement of their meaning and effect.

Oliver Heaviside

. . . formulae . . . the true theoretician does without them as much as possible. He expresses in words whatever he can.

Ludwig Boltzmann

Everything is worthy of notice, for everything can be interpreted.

Herman Hesse

1

Clouds in a Glass of Beer

Mix salt and sand, and it shall puzzle the wisest of men, with his mere natural appliances, to separate all the grains of sand from all the grains of salt; but a shower of rain will effect the same object in ten minutes.
T. H. Huxley

In taverns of low repute the patrons may sometimes be observed to sprinkle salt into their beer. They do this not because they like salty beer but because they are amused by the resulting profusion of bubbles. Yet there is more than amusement in a glass of beer: it exemplifies a surprising number of physical phenomena, many of which occur in the atmosphere, particularly the formation and evolution of clouds. It would not be too much to say that a glass of beer is a cloud inside out.

BUBBLES IN BEER

Beer contains dissolved carbon dioxide (CO_2), molecules of which at a pressure more than twice that of the atmosphere at sea level also occupy the space above the beer in a capped bottle. Although molecules continually flit back and forth between the beer and the gas, the rate at which they leave the beer is balanced by the rate at which they return: CO_2 dissolved in the beer is in equilibrium with that in the neck. In equilibrium the amount of CO_2 dissolved in a given amount of beer (at fixed temperature) is proportional to the pressure of the CO_2 above it. When the bottle is uncapped, this gas escapes rapidly from the neck and its pressure drops greatly. Now the dissolved CO_2 is no longer in equilibrium with that in the neck; the beer is *supersaturated*. In a sense it contains more CO_2 than it ought to, but the excess does not escape explosively: it does so slowly, in the form of small bubbles.

Although many people know about bubbles in beer, the details of their formation sometimes escape notice. A few years ago (1982) an article appeared in *Science* (Vol. 215, p. 1082) entitled "Bubbles upon the River of Time," by M. Mitchell Waldrop. It was about the speculations of Richard Gott, an astrophysicist, that our universe is only one of perhaps an infinite number which formed like bubbles in a very hot dense space. What caught my eye was the following: "Gott's bubbles form just like bubbles in a glass of beer—randomly." This was too good to let pass, so I fired off a letter, which was duly published:

> M. Mitchell Waldrop must not have spent much time in well-lit pubs. For if he had he would not have said about Gott's bubbles that they form just like bubbles in a glass of beer—randomly; they emanate from a small number of definite nucleation sites; cracks in the glass and bits of foreign matter. Moreover, strings of bubbles in which the bubble spacing increases regularly with height above these nucleation sites are easily observed in any glass of light (colored), gaseous American beer.

A bubble consists of gas surrounded by liquid, the two phases separated by a definite surface. It takes energy to form surfaces. For example, when you break a piece of chalk two new surfaces are created. This takes energy, not much, but still a finite amount. The chalk could break spontaneously, but this is highly unlikely. We would have to wait a long time for it to happen, longer than the age of the universe.

Because energy is required to create bubbles in beer they do not form spontaneously under normal conditions: *nucleation* sites, places where bubble embryos can grow in a supersaturated environment, are required. Such tiny invisible bubbles might inhabit microscopic cracks in the glass or in particles on its sides or in the beer.

Salt sprinkled into beer provides a great number of nucleation sites; each grain, which is pitted and cracked, has many such sites. This is shown in Figure 1.1, a photograph of salt grains descending in beer taken by E. Philip Krider at the University of Arizona's Institute of Atmospheric Physics. During a visit of mine to Tucson, Krider interrupted his lightning studies long enough to help me perform an experiment and to enjoy a discussion of the results—over a cold glass of beer, of course.

Although you might think that bubbles evolve from salt grains solely because the grains dissolve in beer, or because of a chemical reaction between beer and salt, this is demonstrably false: merely try insoluble particles. Fine aquarium sand (quartz) is quite suitable. Wash the sand thoroughly in acid, then water, to rid the grain surfaces of soluble impurities. When this clean sand is dropped into beer the resulting bubbles are in no apparent way different in number and size from those nucleated by salt grains, although sand does make the beer a bit gritty.

Figure 1.1 Salt grains descending in beer. Each grain provides many sites for the nucleation of bubbles. Photograph by E. P. Krider.

CLOUD FORMATION

Clouds form when the relative humidity is sufficiently high that atmospheric water vapor, a gas, condenses into small liquid droplets (for more on cloud formation see the following chapter). A cloud is in many ways just the inverse of a bubbly glass of beer. The former is composed of liquid droplets suspended in a gas (air), whereas the latter is composed of gas droplets suspended in a liquid (beer). Cloud droplets fall (unless they are caught in updrafts), whereas bubbles rise. But both owe their existence to nucleation by agents external to themselves. Were it not for the presence of *condensation nuclei,* tiny particles in the atmosphere, clouds would not form under conditions that we have come to accept as normal. Without such particles in the atmosphere, or even ions (which also can serve as condensation nuclei), clouds could still form although it would require much higher relative humidities, perhaps 400 percent or higher. In this instance cloud droplets would form by *homogeneous* nucleation: enough water molecules get together by chance to form stable, long-lived clusters which can then act as nuclei for further condensation. The same is true in beer. If there were no nucleation sites—cracks and so forth—the amount of CO_2 that could be dissolved in beer would be large, but eventually bubbles would form by homogeneous nucleation.

HOMOGENEOUS NUCLEATION IN A BEER BOTTLE

When a bottle of beer is opened, a cloud often forms in the neck (Fig. 1.2), although this may not be noticed by those who frequent poorly lit pubs or who are too eager to guzzle the bottle's contents. The cloud wasn't there when the bottle was capped. Why does it form when it is uncapped?

The space above beer in a capped bottle contains mostly carbon dioxide, but there are other gases and vapors as well. Among them is water vapor because beer is mostly water (some beers seem to be entirely water).

When the bottle is opened, gases and vapors in the neck expand rapidly into the surroundings, as a consequence of which the temperature in the neck drops precipitously. I have estimated that if the temperature is 5°C (41°F) before opening, it drops to about −36°C (−33°F) immediately after opening. The water molecules in the neck become so sluggish that purely by chance several of them can get together long enough to form small embryos that will serve as sites for further condensation of water. This doesn't happen in the atmosphere because it almost always contains particles that may serve as sites for *heterogeneous* nucleation.

There undoubtedly were particles in the space above the beer at the time it was bottled. But this small, enclosed space becomes purged of them. The larger ones settle very quickly. The smaller ones settle much more slowly, but because of their small masses they diffuse rapidly to the sides of the bottle where they

Figure 1.2 Cloud in the neck of a freshly opened bottle of beer. Unlike clouds in the atmosphere, this cloud formed by homogeneous nucleation.

are captured. One way or another the particles are eventually removed. Yet clouds form in the necks of bottles that have been sitting for months, which is surely long enough for nearly all particles to have been removed.

BUBBLE BEHAVIOR

If you carefully observe bubbles in beer you will notice that they are not all the same size. Nor are they evenly spaced. Figure 1.3 is a photograph of strings of bubbles rising from nucleation sites on the side of a glass of beer. Note that the spacing between the bubbles and their size increase more or less steadily with distance from the point of nucleation. This regular spacing suggests that the rate of bubble formation is nearly constant, which in turn implies that a bubble's velocity increases with its size because the spacing and size increase simultaneously. But why does a bubble rise in the first place? And what determines its speed? To answer this we must consider the forces acting on a bubble.

Gravity pulls a bubble downward with a force equal to its weight; buoyancy pushes it upward with a force equal to the weight of beer it displaces. Beer is so much heavier than carbon dioxide that a bubble's weight is negligible compared with the buoyant force on it. A bubble is therefore positively buoyant and rises when it breaks loose from its nucleation site, just like an untethered hot-air balloon. An upward buoyant force is not, however, the only force acting on a bubble: it also experiences *drag*.

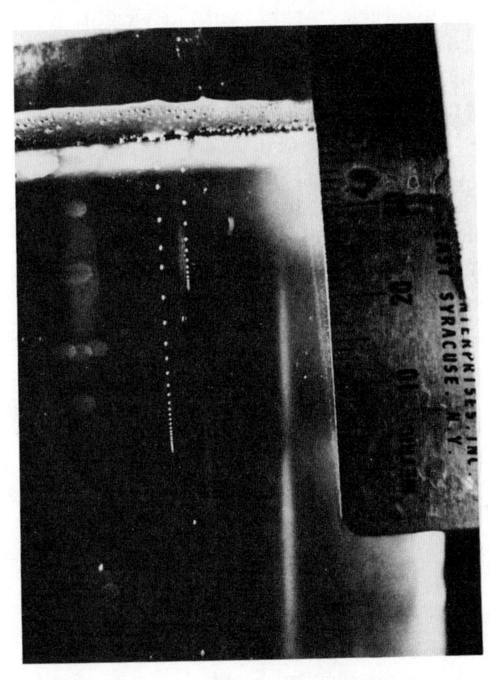

Figure 1.3 Bubbles ascending from nucleation sites on the side of a glass of beer. Note the increasing size and separation of bubbles. Photograph by E. P. Krider.

Your hand, when thrust out the window of a moving car, is hurled backward by the force of wind resistance, or drag. The greater the speed the greater the drag. Bubbles, too, experience a drag force which increases with increasing speed, and this impedes their upward motion. Because the buoyant force, unlike the drag force, does not depend on a bubble's speed, the two forces acting on it must eventually come into balance, and when they do a bubble of *fixed size* moves upward at constant speed.

A bubble released from rest travels only a small fraction of its diameter before reaching a constant speed. Because of the large viscosity of beer (compared with that of CO_2) and the low density of CO_2 (compared with that of beer) the drag and buoyant forces acting on a bubble are nearly always instantaneously in balance. This is true even if the bubble grows suddenly. For a brief moment the two forces are unequal, but in the time it takes the bubble to travel a short distance the forces come into balance again. As dissolved carbon dioxide in the supersaturated beer diffuses into bubbles, these bubbles grow, and their speed increases. A string of bubbles is therefore nonuniform: the greater the distance from the nucleation site the larger the bubble and the larger its separation from adjacent bubbles.

CLOUD DROPLET BEHAVIOR

The same forces acting on bubbles also act on cloud droplets, although with different relative magnitudes. Because water is so much denser than air the upward buoyant force on a droplet is negligible compared with its weight. Also, the drag on a droplet is much less than that on a bubble with the same size and speed because air is much less viscous than beer. The terminal speed of a cloud droplet increases with its size, but it must fall a relatively greater distance—many times its diameter—before attaining this speed. Cloud droplets also grow by diffusion if they are in a supersaturated environment, and the larger they are the faster they fall. You might think that they grow this way until they are large enough and fast enough to become raindrops. But diffusion is a slow process, much too slow to account for the rapidity with which clouds can form and produce rain. So some other mechanism must transform cloud droplets into raindrops. One such mechanism, important in what are called warm clouds (those at temperatures above freezing), is collision and coalescence.

When two cloud droplets collide they sometimes coalesce into a single larger drop. Similarly, bubbles in beer also collide. You can observe this by tilting the glass so that different strings of bubbles intersect. But there the similarity ends. I have never observed them to coalesce, only to bounce.

CLOUD SEEDING AND BEYOND

The connection between beer and clouds sometimes extends into the realm of fantasy, nowhere more so than in a one-act play by E. M. Fournier d'Albe

(*Weather,* July 1960, p. 243) entitled "Why not seed clouds with beer?" Two characters, meteorologists Fulano and Zutano, argue—in convivial surroundings—the merits of seeding clouds with beer. They conclude with the following "new recipe for seeding recalcitrant cumulous clouds":

> Take a sufficient quantity of
> good strong beer;
> Add salt to taste, or until it
> is slightly hygroscopic;
> Add enough soap to give it a
> really good froth;
> Shake well, and administer in
> the form of bubbles to promising
> young cumuli;
> Then stand clear, and try to
> evaluate the results.
> All right?

One cannot help but wonder what would have been the reaction of John Aitken, the Scottish scientist who did so much in the last century to advance understanding of nucleation in the atmosphere, to all this talk of clouds and beer. It seems that a portion of his estate was left to establish a temperance public-house in his native Falkirk. Aitken would no doubt have preferred ginger ale in place of beer. But for several reasons I think that he would have had to grudgingly admit that, taste aside, beer really is superior to ginger ale for making the kinds of observations I have described.

I have by no means exhausted all the physics in a glass of beer. For example, I have said nothing about why bubbles migrate across the surface of the beer to the edge of the glass and accumulate there in rafts. And if you whip the beer into a froth the resulting head is, like a cloud, white, although the liquid from which the bubbles in the head were formed is yellow (for more on this see Chapter 15). And bubbles shrink as they near the surface, which has been observed by alert beer physicists. Investigation of these important matters is best left for a hot day in July.

GREAT MINDS THINK ALIKE: A POSTSCRIPT

Soon after the article on which this chapter is based first appeared in *Weatherwise* (October 1981), Jearl Walker published something remarkably similar in *Scientific American* (December 1981). Yet our articles were completely independent. I certainly did not know about his, and he learned of mine only after his had gone to press. Beer physics must have been in the air in the fall of 1981, and the two of us merely happened to be the lucky ones who plucked it out.

2

Genies in Jars, Clouds in Bottles, and a Bucket with a Hole in It

But presently there came forth from the jar a smoke which spired heavenwards into ether . . . and which trailed along earth's surface till presently, having reached its full height, the thick vapor condensed, and became an Ifrit [genie, also spelled jinni].

The passage at the head of this chapter is from "The Fisherman and the Jinni," one of the tales from the *Arabian Nights*. When the fisherman in this tale uncorked the jar, its contents expanded rapidly into the surroundings, and the vapor it contained condensed into a genie. So also does water vapor in expanding air condense into cloud droplets.

A CLOUD IN A BOTTLE

I am unable, alas, to conjure a genie from a jar. But I can make a cloud in a bottle, and so can you. I use a large bottle, half of which is painted black so that the cloud will be more noticeable when illuminated by a bright light. The bottle should have a stopper with a hole in it. Put a little water in the bottle, just enough to cover its bottom, and blow hard into a length of tubing inserted through the hole in the stopper. Seal the end of the tubing with your finger, then release it suddenly. The result is likely to be disappointing (Fig. 2.1) because I have forgotten something: more particles are needed. Since the bottle is an enclosed space, many of the particles in it have either settled out or have diffused to the walls, especially if it has been stoppered for a long time (see the previous

Figure 2.1 An unsuccessful attempt at making a cloud in a bottle. Photograph by Gail Brown.

chapter). Particles are provided readily enough by a smoldering match. First decrease the pressure in the bottle by sucking out some of the air. Wave the match near the end of the tubing; as air rushes back into the bottle, it will carry with it some of the smoke. Now try once again to make a cloud. This time your efforts are more likely to be met with success (Fig. 2.2); if not, try adding more particles. To understand why the cloud forms and what the particles have to do with it, I must first discuss a few concepts and elucidate them with further demonstrations.

Figure 2.2 A successful attempt at making a cloud in a bottle. Photograph by Gail Brown.

SATURATION VAPOR PRESSURE

Let us consider a hypothetical experiment; we won't actually do it, we'll just imagine it to be done. Take a bottle, partly filled with water, and cork it. But before corking it, remove all the water molecules from the space above the liquid surface. This space will not, however, remain free of water molecules for long. Molecules in the liquid are continually jostling about and colliding with one another. Every now and then a molecule will acquire a bit of extra energy from its neighbors, sufficient to allow it to overcome their attraction, and will escape into the space above the liquid. This will occur again and again at a very rapid rate. As the number of water molecules—water in the gas phase—in this space increases, the rate at which they return to the liquid also increases. Eventually, the rate at which water molecules leave the liquid phase and enter the gas phase—the rate of *evaporation*—is balanced by the rate at which the reverse process—*condensation*—occurs. Thus a dynamic equilibrium exists: the level of the liquid remains constant as does the amount of water vapor in the space above it, although there is a continuous exchange of molecules between the liquid and gas phases. When equilibrium is reached, the *partial* pressure of the water vapor (small compared with the *total* pressure, the sum of partial pressures contributed by each of the constituents of air) is called the *saturation vapor pressure*. But equilibrium vapor pressure would be a better term: "saturation" evokes, incorrectly, the image of a sponge. There is no end of blather about the "holding power of air" and how air can "hold" more water vapor at high temperatures than at low temperatures; this implies that in air there is only so much space—like rooms in a hotel—between air molecules, and when filled with water molecules the air is saturated, just like the pores of a sponge. But air doesn't "hold" water vapor—it coexists with it. Indeed, the presence of air (oxygen, nitrogen, etc.) in the space above the liquid is largely immaterial: the pressure of a vapor in equilibrium with its liquid would be nearly the same with or without air. It is worth noting here that *everything* has a vapor pressure; that of mercury at room temperature, for example, is about one-thousandth that of the atmosphere at sea level. That of most solids, especially near room temperature, is very much lower—but it is not zero. Your skin has a vapor pressure. Fortunately, it is rather low or you wouldn't be here to read this—you would have evaporated away long ago.

If the notion that air "holds" water vapor were correct it would necessarily follow that the saturation vapor pressure would increase if the distance between air molecules were increased by reducing the air density, thereby providing more room for water molecules. But the saturation vapor pressure above a flat surface of pure water depends only on temperature (this will be qualified ever so slightly in Chapter 4), and it increases with increasing temperature. Why this is so is easy to understand. The greater the temperature, the greater the energy of the molecules in the liquid and the easier it is for them to escape. And the greater the rate of evaporation, the greater will be the vapor pressure once equilibrium is reached. A bucket with a hole in it helps to explain why by analogy.

A BUCKET WITH A HOLE IN IT

Make a hole near the base of an empty bucket (a plastic bottle will serve just as well); then pour water into it, from a tap for example, at a constant rate. Any liquid would do, water is just the handiest. Initially there is no water in the bucket so none can leak out. As the water level rises, however, the rate at which water leaks from the hole will increase. Eventually, the water will leak out as fast as it pours in, and the water level will be constant—dynamic equilibrium has been reached. Now increase the rate of inflow—give the tap a twist. The water level will rise to a new, higher equilibrium level. The height of water in the bucket once equilibrium is reached is analogous to the saturation vapor pressure above a liquid (or a solid for that matter); the constant rate of inflow is analogous to the rate of evaporation of the liquid; and the rate at which water leaks out of the hole is analogous to the rate of condensation of vapor. Moreover, the equilibrium height of water in the bucket depends only on the rate of inflow. By analogy, therefore, the saturation vapor pressure depends only on the rate of evaporation, which in turn depends only on the temperature. So the saturation vapor pressure is in one sense merely a measure of the rate of evaporation, and it is often advantageous to look at it this way.

HOW DOES A CLOUD DROPLET FORM?

Two qualifications crept quietly into the previous discussion: I stated in passing that the saturation vapor pressure above a *flat surface of pure water* depends on temperature only. What if the surface is not flat? Or the water not pure?

The rate of evaporation from a water droplet increases with decreasing radius because a water molecule at the surface has fewer neighbors attracting it. So the smaller the droplet the easier it is for a water molecule on its surface to escape. As far as evaporation is concerned, droplets larger than about 1 μm (a millionth of a meter) are flat surfaces. But the rate of evaporation from a droplet of radius 0.001 μm, for example, is more than three times that from a droplet of radius 1 μm, both at the same temperature. Just as every general was once a lieutenant, and every frog once a tadpole, every cloud droplet of radius 10 μm, say, was once much smaller: cloud droplets grow, they aren't hatched fully fledged. This presents us with a puzzle. For a droplet to grow it must be in an environment such that the rate of condensation is greater than the rate of evaporation, which makes rather severe demands on the environment.

Relative humidity is usually defined as the actual vapor pressure of water divided by its saturation vapor pressure above a *flat surface of pure water* at a given temperature. Another definition, one which I much prefer because it incorporates physical processes, is that it is the rate of condensation relative to the rate of evaporation. If a cloud droplet of radius 0.001 μm is not to shrink, the relative humidity of its environment must be greater than 300 percent because the rate of evaporation from such a small droplet is about three times that from a flat surface. Such relative humidities are not observed in the atmosphere. I do not doubt the existence of clouds, they are a matter of everyday experience; yet I have just presented an argument that calls their existence into question. There must be more to the formation of clouds than I have yet revealed.

A DEMONSTRATION WITH SALT GRAINS

Cloud droplets can avoid being very small by beginning their existence as solid particles, or *condensation nuclei* (see the previous chapter). If water vapor condenses on a nucleus of radius 0.1 μm, say, growth is possible in an environment with a modest amount of supersaturation (relative humidity in excess of 100 percent), a few percent or less. And if the condensation nucleus is soluble, so much the better: the saturation vapor pressure above a solution is less than that above pure water, both at the same temperature (the reason for this is discussed in Chapter 4). Thus a salty water droplet will evaporate at a slower rate than a pure water droplet, both of which are the same size and at the same temperature. This may be demonstrated with grains of table salt even though they are enormous compared with particles in the atmosphere.

Sprinkle a few salt grains onto the lid of a tin can elevated above the bottom of a dish (Fig. 2.3); bent paper clips serve nicely as supports for the lid. For classroom demonstrations I put the salt grains on a glass slide in a Petri dish, a transparent glass dish used for growing bacterial cultures, and project their image onto a screen as far from the projector as possible to obtain the greatest magnification. Note in Figure 2.3 that the grains are mostly cubes. Now carefully add some water to the dish, just enough to wet the bottom, and then cover it. If you wait a bit, perhaps twenty minutes or more, the cubes will be transformed into hemispheres (Fig. 2.4). Water evaporates from the liquid and condenses onto the salt grains; in so doing it dissolves them. Of course, water also

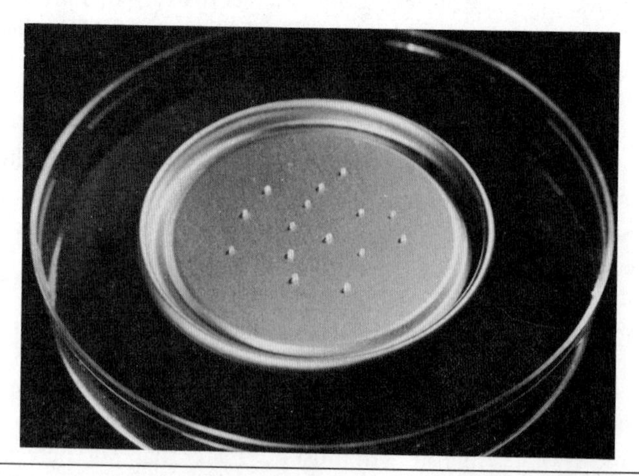

Figure 2.3 Dry salt grains elevated above the bottom of a dish.

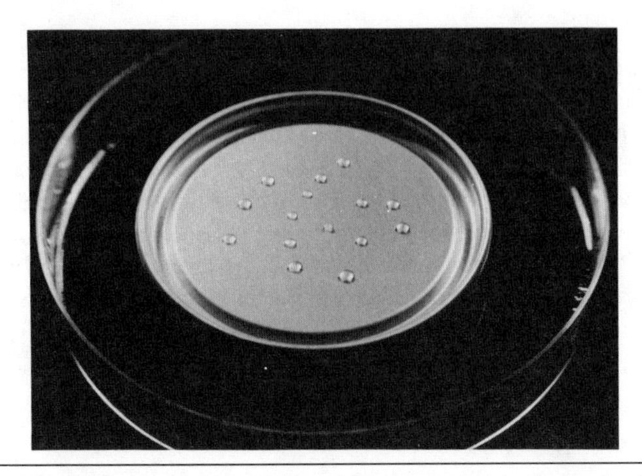

Figure 2.4 The grains of Figure 2.3 are transformed into salty water droplets when a little water is put in the dish and it is covered.

evaporates from the grains, but at a slower rate than from the pure water. Hence there is a net transfer of water from the bottom of the dish to the grains. Given enough time each grain will completely dissolve and in its place will be a droplet of salty water. A very thin film of oil on the surface holding the salt grains results in more pronounced droplets (see Chapter 7). The grains accumulate water only in an environment with a high relative humidity, about 80 percent or higher. Unless you are in a very humid room, therefore, the droplets will evaporate when the disk is uncovered, each leaving in its wake a residue of salt (Fig. 2.5).

Salt particles are just one among many types of condensation nuclei. How does salt get into the atmosphere? When students are asked this question, the

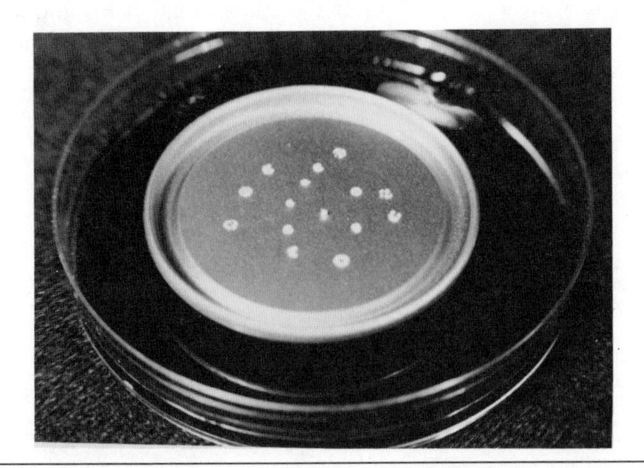

Figure 2.5 The water droplets in Figure 2.4 evaporate when the dish is uncovered leaving behind a residue of salt.

usual response is evaporation of ocean water. Yet, as we have seen, when salty water evaporates, the salt is left behind; it doesn't evaporate into the air. The mechanism by which salt gets into the atmosphere forms one of the more interesting chapters in cloud physics, and a fascinating discussion of it is given in Duncan Blanchard's very readable book *From Raindrops to Volcanoes: Adventures with Sea Surface Meteorology.* In brief, when bubbles of entrained air break through the ocean surface small droplets of salt water are formed; some of these are carried away by updrafts and evaporate; this salt residue, in the form of small particles, can then serve as condensation nuclei for clouds far away in space and time.

BACK TO THE CLOUD BOTTLE

At last we have done enough to explain the first demonstration. After increasing the total pressure in the bottle by blowing into it, and then suddenly releasing the pressure, a visible cloud was formed—if there were enough particles in the bottle. When air rushes out of the bottle, the air left behind cools rapidly. Cooling upon expansion is familiar to anyone who has ever noticed a pressurized spray can cool; sometimes ice even forms on the can. As the air in the bottle cools, the rate of evaporation of water from the particles decreases. The rate of condensation also decreases because water vapor was lost from the bottle (this water vapor would be replenished from the liquid water in the bottle in a minute or more, but the cloud forms in a much shorter time). Evaporation depends so strongly on temperature, however, that the decrease in evaporation exceeds that in condensation. Condensation therefore exceeds evaporation, and the nuclei grow rapidly into droplets large enough to strongly scatter light thereby making their presence visible.

I leave it to your ingenuity to put the proper ingredients—liquids, solids, and particles—into a jar to ensure that a genie condenses out of the gas phase when the jar is uncorked.

3

Happy Ducks, Like Happy People, Perform Best with Cool Heads

To play billiards well is a sign of a mis-spent youth.
Herbert Spencer

A trinket likely to be found in the kinds of gift shops that infest seaside resorts is the toy duck shown in Figure 3.1. It goes by various names; mine is called the Happy Duck. Sometimes such a duck may be seen in a shop window, endlessly dipping its beak in water placed before it (Fig. 3.2). Any inquisitive person naturally wants to know what impels this ever-thirsty bird.

MECHANICS OF THE HAPPY DUCK

Except for its head and a solitary tailfeather, the duck shown in Figures 3.1 and 3.2 has been denuded—the better to see how it works. The duck is made up of two hollow glass bulbs, its head and its bottom, connected by a glass tube which penetrates almost all the way into the bottom bulb (Fig. 3.2). A volatile liquid partly fills the bottom bulb and the tube to the same level when the duck is upright (Fig. 3.1). If you wet the duck's head, which is covered with an absorbent material, the liquid in the tube will rise slowly—unless the humidity is very high. If you are impatient you can speed things up by blowing in the duck's ear. And if you really want fast action you can grab its bottom.

The duck is free to rotate about a horizontal axis through its legs. As the liquid rises in the tube, the duck becomes top-heavy and its head swings forward, its arc limited by small stops on its legs. When it reaches the end of its arc, with its body almost horizontal, the tip of the tube is above the liquid level

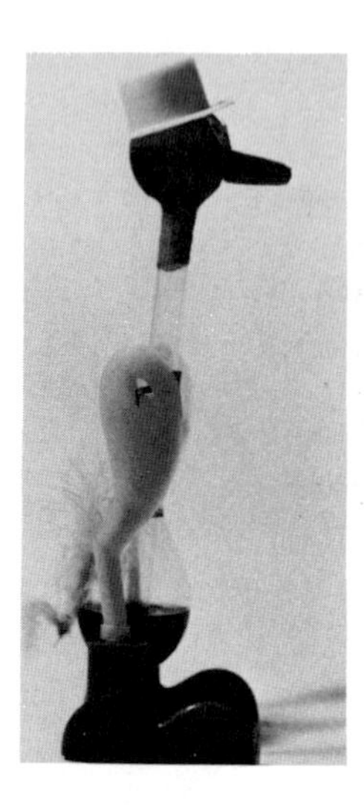

Figure 3.1 The Happy Duck. Photograph by Gail Brown.

and bubbles rise in the tube; these can be seen in Figure 3.2. Then the liquid in the tube flows back into the bottom bulb, and the duck returns to the upright position (Fig. 3.1). In a moment the liquid rises in the tube again and the cycle is repeated. If the duck's head is wet and its bottom dry this cycle will be repeated indefinitely.

What I have just described is easy enough to observe. The only mystery to be explained is why the liquid rises.

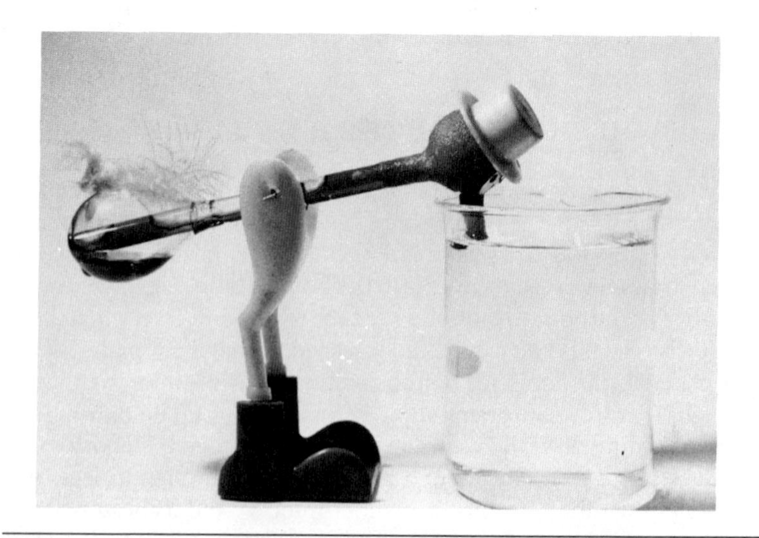

Figure 3.2 The Happy Duck takes a drink. Note the bubbles rising in the tube. Also note that the duck's head is wet. Photograph by Gail Brown.

EVAPORATIVE COOLING

The liquid rises in the tube because the duck's head, if it is wet, is colder than its bottom because of evaporative cooling. In the previous chapter I discussed evaporation but did not emphasize that it is a cooling mechanism. I merely stated that "Every now and then a molecule will acquire a bit of extra energy from its neighbors, sufficient to allow it to overcome their attraction, and will escape into the space above the liquid." So it is the more energetic ones that escape, leaving behind the less energetic ones. This, in turn, is accompanied by a *decrease* in temperature of the liquid. Condensation, the reverse process, must therefore result in a temperature increase. Sir James Jeans made an analogy which I have adapted to help clarify why molecules leaving the liquid result in cooling whereas those returning result in warming.

Consider a billiard table on which there are many balls whizzing about ceaselessly. It is a peculiar table because it is divided into two parts, one lower than the other, connected by a ramp. On the low side of the ramp there are balls, on the high side none—at first. Occasionally a ball approaches the ramp, and if it has enough kinetic energy (energy of motion) it can get over; if not, it rolls back. So the ramp acts as a kind of filter, analogous to the liquid-vapor interface, which allows only the energetic balls to escape the region of madly whizzing balls; this is analogous to evaporation. Balls whizzing about on the high side of the ramp may, of course, approach the ramp and roll down it. When they do, they acquire kinetic energy; this is analogous to condensation. If evaporation exceeds condensation (i.e., there is *net* evaporation), this leads to cooling. When condensation and evaporation are equal the liquid neither cools nor warms. This is what happens when the relative humidity is 100 percent, which is why I made the qualifying statement about the duck not working on very humid days. For net evaporation to occur, the vapor pressure must decrease outward from the liquid surface, and the greater this decrease (i.e., the greater the *vapor pressure gradient*), the greater the rate of evaporation. This is why I recommended blowing on the duck. Air moving past its head whisks away the water vapor molecules, thereby maintaining a higher vapor pressure gradient than would otherwise prevail. I also recommended grabbing the duck's bottom to hasten the rising liquid. The liquid rises because the duck's head is colder than its bottom, and this can be achieved either by cooling its head by evaporation or by warming its bottom with your hand. Now all I have to do is explain why this temperature difference causes the liquid to rise in the tube.

I emphasized in the previous chapter that everything has a vapor pressure, which includes the liquid in the duck. And the vapor pressure of this liquid, like that of water, increases rapidly with increasing temperature. Regions inside the duck not occupied by the liquid are occupied by molecules of its vapor. When the duck's temperature is the same everywhere so is the pressure exerted by this vapor. But when its head is colder than its bottom the vapor pressure in its head is less than that in its bottom. It is this vapor pressure difference which causes the liquid to rise. The liquid falls when the tip of the tube is uncovered and bubbles rise in it thereby equalizing the pressure in the two bulbs. Although the tube is a kind of barometer, the pressure differences it responds to are *inside* the duck. What the pressure is in the surrounding atmosphere is largely immaterial. The only atmospheric variable important to the smooth working of the duck is relative humidity, which suggests a way to measure relative humidity.

DUCK-BOTTOM AND DUCK-HEAD TEMPERATURES

When the relative humidity is 100 percent the temperature of the duck's head, even if it is wet, is the same as that of its bottom. In general these two temperatures will be different. And the greater this temperature difference the smaller the relative humidity. As the relative humidity decreases the vapor pressure gradient near the duck's head increases and so, therefore, does the rate of evaporative cooling. Thus the temperature difference between the duck's wet head and its dry bottom is a measure of the relative humidity. These two temperatures, the duck-head and the duck-bottom temperatures, are usually given the more prosaic names *wet-bulb* and *dry-bulb* temperatures, respectively. With them you can determine the relative humidity.

It is a bit difficult to attach thermometers to the duck. An equivalent procedure is to measure the wet-bulb and dry-bulb temperatures using an inex-

Table 3.1 Relative humidities (%) for various dry-bulb temperatures and wet-bulb depressions (dry-bulb – wet-bulb).

Dry-bulb	Wet-bulb depression (°F)														
(°F)	2	4	6	8	10	12	14	16	18	20	22	24	26	28	30
35	81	63	45	27	10										
40	83	68	52	37	22	7									
45	86	71	57	44	31	18	6								
50	87	74	61	49	38	27	16	5							
55	88	76	65	54	43	33	23	14	5						
60	89	78	68	58	48	39	30	21	13	5					
65	90	80	70	61	52	44	35	27	20	12	5				
70	90	81	72	64	55	48	40	33	25	19	12	6			
75	91	82	74	66	58	51	44	37	30	24	18	12	7	1	
80	91	83	75	68	61	54	47	41	35	29	23	18	12	7	3
85	92	84	77	70	63	56	50	44	38	33	27	22	17	13	8
90	92	85	78	71	65	58	52	47	41	36	31	26	22	17	13
95	93	86	79	73	66	60	55	50	44	39	34	29	25	21	17

pensive aquarium thermometer and a piece of old undershirt. Such thermometers are often mounted on a stiff piece of plastic to which you can attach a length of string. First measure the temperature as you would ordinarily; this is the dry-bulb temperature. Then wrap a small piece of absorbent cloth around the tip of the thermometer; the cloth can be held in place with a rubber band. After thoroughly wetting the cloth, whirl the thermometer rapidly by the string. Evaporative cooling will cause the temperature to decrease. Keep whirling the thermometer and every so often check the temperature it indicates. The lowest temperature thus noted is the wet-bulb temperature. In addition to these two temperatures, wet-bulb and dry-bulb, psychrometric tables or charts are needed to obtain the relative humidity. I can't give you a complete set of psychrometric data, but Table 3.1 has enough values of dry-bulb temperatures and wet-bulb depressions (difference between the dry-bulb and wet-bulb temperatures) to enable you to estimate the relative humidity.

A much fancier—and costlier—device for measuring relative humidity is called a sling psychrometer: two thermometers, one for measuring the wet-bulb temperature, the other for measuring the dry-bulb temperature, mounted together on a stand which can be rotated rapidly by hand. But my cheap sling psychrometer works about as well as much more expensive ones.

To obtain accurate results with a psychrometer, one must wet the wicking with pure water. Why this is so is the subject of the next chapter.

4

Sugar and Spice: The Dirty Wet-Bulb Temperature

*Nine times out of ten . . . there is actually
no truth to be discovered . . .
only error to be exposed.*
H. L. Mencken

Gail Brown and I once wrote disparagingly about the fatuous slogan that cold air can't hold as much water vapor as warm air (see Chapter 2). In response to our article we received a letter from Duncan Blanchard, a professor of atmospheric sciences at the State University of New York at Albany. He agreed that our point was a good one, but feared that "the idea of air holding water will keep coming up like dandelions in the spring." This is an apt simile, although I would have chosen a more obnoxious and robust weed. Perhaps kudzu would be more appropriate.

For brevity, I henceforth shall refer to the slogan that cold air can't hold, etc., as the windbag argument.

Unfortunately, this argument has become part of the oral and written tradition of meteorology. It appears in print with dismal regularity. The windbag argument, fatuous though it may be, is in itself harmless. It is a thought-cliche, that is, an avoidance of thought, but it falls short of being truly idiotic. What is idiotic are the attempts to explain *why* warm air holds more water vapor than cold air.

I have seen it stated that as air is heated it expands (true enough, unless it is confined), the separation between air molecules therefore increases (also true), thus the air can hold more water vapor (piffle). Sometimes an analogy is made to a hotel: once all the rooms are occupied, guests are turned away. To make this analogy is to defile the grave of John Dalton, to whom science in general, and meteorology in particular, owes so much. For it was Dalton, a Manchester schoolteacher, who announced in 1802 his conclusion, based on experiment, that the vapor pressure of water in air is independent of the existence of the air; the same results are obtained in a vacuum.

Despite this, the windbag argument flourishes like kudzu. The time has come to apply some weed killer in the form of a simple demonstration and its interpretation. This demonstration arose from conversations with Dennis Thomson, one of my colleagues at Penn State.

SUGAR AND SPICE

In the previous chapter, I discussed the concept of wet-bulb temperature and how to measure it. Now I suggest that you redo the experiment I described, but this time use concentrated solutions instead of pure water to wet the wicking.

First measure the wet-bulb temperature in the usual way. Then repeat the measurement after wetting the wicking with water in which you have dissolved some ordinary table salt. A highly concentrated solution gives the best results, so pour in plenty of salt. You will find that the wet-bulb temperature is higher, perhaps a few degrees. Now measure the wet-bulb temperature again using new wicking that has been dipped into a sugar solution. Again, you will want a highly concentrated solution, so don't spare the sugar. You don't have to prepare gallons of solution, just enough to wet the wicking. This wet-bulb temperature will be closer to, but still greater than, that obtained with clean wicking. All this is simple enough to do. What is more difficult is to explain the results.

Wet-bulb temperatures are lower than dry-bulb temperatures (except if the relative humidity is close to 100 percent) because of evaporative cooling, which was explained in the previous chapter; the greater the rate of evaporation of water, the lower the wet-bulb temperature. To be more precise, the greater the *difference* between the evaporation and condensation rates, the greater the *difference* between the wet-bulb and dry-bulb temperatures. For each measurement the rate of condensation of water vapor onto the wicking is the same because this is determined by the concentration of water vapor in and the temperature of its surroundings. We therefore conclude that dissolving something in water lowers its rate of evaporation.

In other words, the saturation vapor pressure above a solution is less than that above pure water, both of which are at the same temperature (as you may recall, I have emphasized that saturation vapor pressure is a measure of evaporation rate). Proponents of the windbag argument, to be consistent, would therefore have to conclude that air above a solution can't hold as much water vapor as air above pure water. Water molecules that have escaped from solution have been tainted somehow by associating with foreigners. The air senses this, and being haughty stuff, refuses to admit so many of them. Air is not only a hotel, it is a highly exclusive one at that. This, I suppose, is molecular blackballing.

Yet the water molecules that escape from solutions are no different from those that escape from pure water. The air plays a passive role; without it, the saturation vapor pressure would be (almost) the same. Thus it is not to the atmospheric repository of escaped water molecules that we must look for an explanation of the experiment, but rather to their place of origin.

RAOULT'S LAW

In 1887, François Raoult, a chemistry professor at Grenoble, published a brief paper in which he announced a law that bears his name today. Although he had predecessors who recognized that adding a solute, such as salt, to a solvent, such as water, lowers the vapor pressure of the solvent, he was the first to cast this in the form of a "general law of the vapor pressure of solvents." I have paraphrased Raoult's version of his law as follows: the reduction of vapor pressure of a solution relative to the vapor pressure of the pure solvent is nearly equal to the ratio of the number of solute to solvent molecules. The form of Raoult's law usually given in modern textbooks is slightly different: the relative

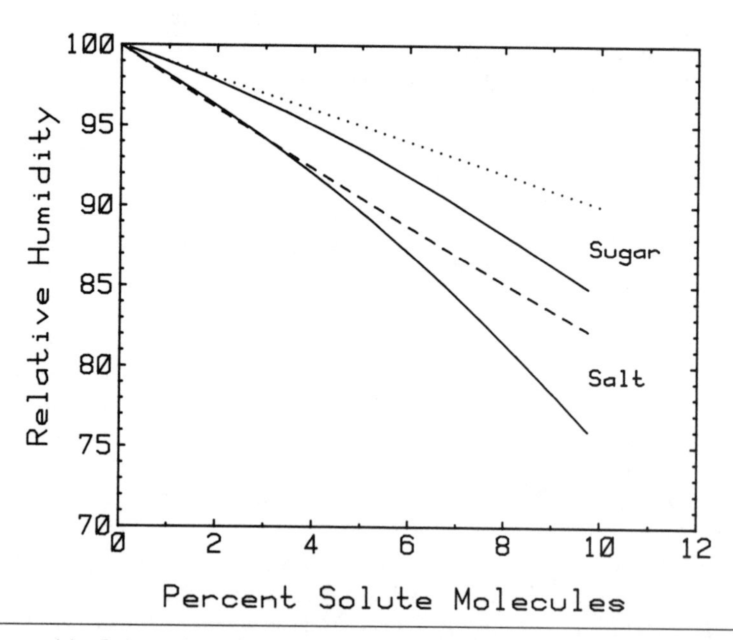

Figure 4.1 Relative humidities above solutions are less than those above the corresponding pure solvent, both at the same temperature. If Raoult's law were strictly valid, the vapor pressure decrease would be proportional to the fraction of dissolved molecules (dotted and dashed lines). But because of interactions between the solute and solvent molecules, the actual decrease (solid lines) is greater. Salt (sodium chloride) is more effective at reducing vapor pressure than sugar (sucrose) because salt dissociates whereas sugar does not.

reduction in vapor pressure is equal to the ratio of the number of solute molecules to the *total* number of molecules in the solution. Regardless of its form, Raoult's law is obeyed only by *ideal* solutions, which, like so many other ideals, exist only in our minds, not in the real world.

Raoult himself recognized that his law was not exact; in his own words, the "proportionality is seldom rigid." This is evident from Figure 4.1, which shows the relative humidity (i.e., the saturation or equilibrium vapor pressure relative to what it would be above pure water at the same temperature) above sugar and salt solutions of varying concentration. If Raoult's law were exact these curves would be straight lines for all concentrations, but they are straight only for low concentrations.

The physical interpretation of Raoult's law is as follows. Dissolving anything in water (for example) decreases the concentration of water molecules, hence fewer of them can leave a unit area of the surface in unit time (e.g., a square centimeter per second). Thus the saturation vapor pressure decreases because it is a measure of the evaporation rate (I keep saying this because it is the key to understanding saturation vapor pressures). This interpretation, which is illustrated in Figure 4.2, is more or less forced on us by the observation that it is not *what* is dissolved (provided the solvent and solute molecules are not vastly different) but merely *how much* which determines the lowering of the vapor pressure (at least at low concentrations). Raoult's law and the interpretation of it I have given are more closely followed the greater the similarity between the solute and solvent molecules.

Now we are better equipped to understand why sugar and salt solutions exhibit the differences uncovered in the demonstration and shown in Figure 4.1. Note that (at low concentrations) each salt molecule dissolved is roughly twice as effective at lowering the vapor pressure as each sugar molecule. I say "salt molecule" even though no such entity exists. A grain of sodium chloride is a regular array (i.e., a crystal) of sodium and chlorine atoms. There is no sodium chloride molecule consisting of a single sodium atom and a single chlorine atom

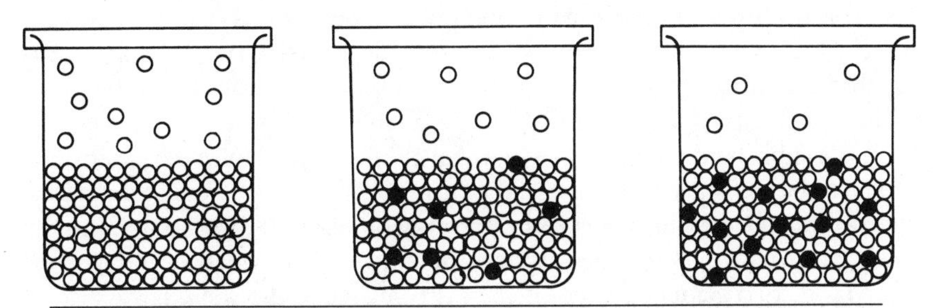

Figure 4.2 Molecular interpretation of Figure 4.1. The rate of evaporation of a liquid, hence the saturation vapor pressure above it, depends on the concentration of evaporating molecules (open circles). As more solute molecules (solid circles) are dissolved in the liquid, the concentration of evaporating molecules decreases, hence so does the saturation vapor pressure.

bound together in a stable unit, but rather the entire crystal should be looked upon as a single giant molecule. So what I mean by a salt molecule is simply a sodium-chlorine pair. The salt is twice as effective at lowering the vapor pressure because each salt molecule dissolved gives rise to *two* ions—positively charged sodium and negatively charged chlorine—whereas sugar in solution does not dissociate into ions.

Deviations from Raoult's law occur when the substance dissolved becomes important, not merely how much of it (i.e., how many molecules). At low concentrations the average environment of a water molecule is almost what it would be in pure water, but as the amount of solute is increased this is no longer so. For example, ions may be hydrated, that is, they form associations with several water molecules, and the vapor pressure is lowered more than would be expected solely from considerations of concentration. Each substance dissolved behaves somewhat differently, so generalizations are risky. All that can be said is that the vapor pressure decreases because of reduction in concentration of the evaporating molecules and because they interact with the solute molecules, the latter becoming increasingly important as the amount of dissolved substance is increased.

DISSOCIATION AND SCIENTIFIC CONSERVATISM

It is easy enough to state dogmatically that when salt is dissolved in water, the result is positively charged sodium ions and negatively charged chlorine ions swimming about in water. Yet this notion was stoutly resisted when it was first put forward in 1883 by Svante Arrhenius in his doctoral dissertation submitted to the University of Uppsala in Sweden. He had an uphill battle getting his ideas accepted, and for good reasons. Sodium is a highly reactive metal. If you can get your hands—oops, don't touch it—on a lump of it, carve off a small piece, and throw it into water, you'll get a vivid demonstration of what is meant by reactive. Chlorine gas is yellowish-green, poisonous, and also highly reactive. It is not used to purify swimming pools because of its benignity toward microorganisms. And yet there was Arrhenius, an audacious young pup, asserting that a solution of sodium chloride, something we wouldn't hesitate to swallow, consists of sodium and chlorine swimming in water.

This is a bit hard to swallow when you think about it. Yet Arrhenius was proven to be right, partly because of the efforts of Wilhelm Ostwald (who nevertheless didn't believe in atoms and vigorously opposed those who did until he finally backed down under the weight of incontrovertible evidence) and Jacobus van't Hoff, two giants of physical chemistry. Chemically, there is a world of difference between the sodium atom—uncharged—and one from which an electron has been stripped—the sodium *ion*. And similarly for the chlorine atom and the chlorine ion. This is why a solution of sodium and chlorine is so benign, whereas each un-ionized component is highly reactive.

The story of Arrhenius could be cast in such a way that he was a hero and his foot-dragging detractors were villains. I must say, however, that I am not opposed to scientific conservatism. Indeed, it is necessary (although when faced with it myself I chafe and writhe and say bad words). We forget that many cockeyed ideas that were resisted by the savants of the day—the Establishment is the pejorative term used—are often shown to have been—cockeyed. Every now and then a rare genius turns out to have had a good idea despite initial resistance to it. And subsequently, hordes of crackpots try to make capital out of this: Arrhenius was ridiculed, he was right; I am ridiculed, therefore, I, too, am right. A manifestly faulty syllogism, but one widely appealed to nevertheless.

Arrhenius didn't have to wander long in the wilderness without recognition. In due time, every scientific honor was heaped on him. Learned societies rushed to elect him to membership, and he was festooned with medals like a military dictator. If you want to read more about this chapter in the history of chemistry, I recommend Bernard Jaffe's book *Crucibles: The Story of Chemistry*. And don't stop with the chapter on Arrhenius, read the entire book. It is exceptionally well written, a masterpiece of popular scientific exposition.

A LESSON FOR HOWARD JOHNSON

It is time for more weed killer, an even stronger dose. Let me further refute the notion that air holds more water vapor the hotter it is because the separation between molecules increases. A necessary corollary to this is that if the separation between molecules in air is decreased by increasing its pressure, then the saturation vapor pressure must decrease (all rooms filled, sorry, go somewhere else). But the reverse occurs: by making fewer rooms, more occupants can be held (a trick that Howard Johnson would have loved to master). I shall try to explain why.

For the sake of easy visualization imagine a closed container, partly filled with pure water, into which air can be admitted. Suppose that we increase the pressure of the air in this container while keeping the temperature constant. What happens to the partial pressure of the water vapor?

Some of the air (eventually) dissolves in the water. As we have seen, dissolving anything in water *lowers* its saturation vapor pressure. But this is not all that happens.

As the total pressure of the liquid water increases so does the concentration of its molecules, that is, the average distance between them decreases. In the previous section, I argued that adding a solute to water *decreases* the concentration of water molecules, hence their evaporation rate *decreases*. It follows from this that if the concentration of water molecules *increases* their rate of evaporation *increases*.

But the story is not complete. Two factors determine the evaporation rate: the concentration of molecules in the liquid and the average energy required to remove a molecule from it. I have considered only the first one. To understand the second, we must examine carefully how the force between molecules depends on the distance between them.

In a gas, molecules are so far apart on average that they are hardly aware of one another's presence; they attract one another only very weakly except during the comparatively short time intervals of collisions. In liquids and solids, however, the molecules are much closer together, and consequently can exert strong forces on one another, which, depending on the distance between them, can be either attractive or repulsive. You encounter examples of this every day, probably without realizing it. For example, take a rubber cork and squeeze it; the more you compress it, the more difficult it is to compress further. When the molecules are close together they *repel*. But now try to pull the cork apart. Again, the molecules resist: when their separation is increased, they *attract* one another. This is shown in Figure 4.3.

In this same figure I show the energy required to take two molecules initially a specified distance apart and separate them by a very large distance (strictly speaking, infinitely large). Negative energies just mean that one must supply energy to separate the molecules; positive energies mean that energy is released upon separation. An analogy may help to explain this. To remove a boulder from a ditch requires energy. You have to lift it (i.e., exert a force) a certain distance; the shallower the ditch, the less energy is required. But if the boulder is outside the ditch on ground sloping downward, you can give it a slight kick and it will roll downhill without any exertion on your part.

Although a molecule *within* a liquid is subjected to both attractive and repulsive forces, the *net* force on it is zero, on average, because it is attracted (and repelled) in opposite directions by surrounding molecules. So molecules

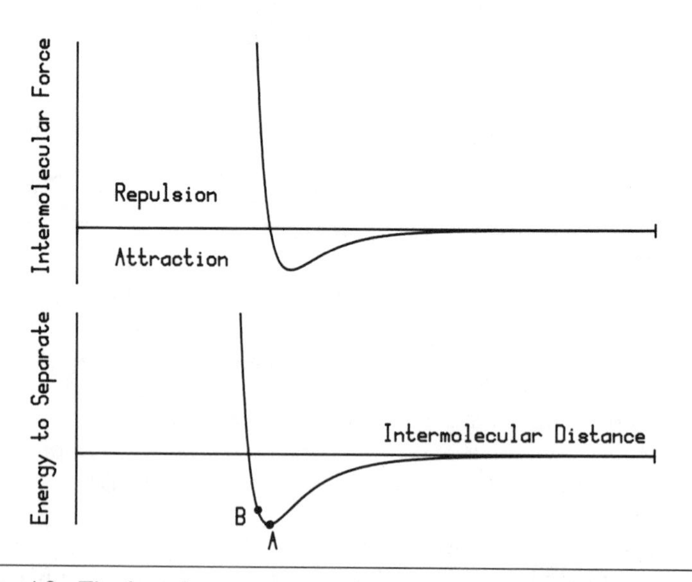

Figure 4.3 The force between two molecules that do not combine chemically is either attractive or repulsive depending on the distance between them (top). The energy required to completely separate two molecules is therefore either positive or negative (bottom).

of all energies can move about freely within the liquid. But as a molecule near the surface escapes the liquid it experiences a net attractive force inward (there are fewer molecules above the surface to attract it outward) and a smaller net repulsive force outward. Suppose that the average distance between a molecule near the surface and its neighbors is about that shown by the point A in Figure 4.3. Now imagine that we push the molecules in the liquid closer together so that this distance decreases (point B). The energy that a molecule must have in order to escape also decreases. This is what happens when the pressure on water is increased: the average distance between water molecules decreases, hence it takes less energy to remove one from the liquid, and the evaporation rate therefore increases. Another way of describing this is as follows. An escaping molecule runs a gauntlet of attractive forces which hinder it and repulsive forces which help it. It is hindered more than helped, and only molecules with sufficient energy escape. The more densely packed the molecules, the weaker the attractive force and the stronger the repulsive force.

The increase in evaporation with increased pressure of the water is much greater than the decrease caused by air dissolved in it. But the story is still not complete.

I have repeatedly emphasized that the saturation vapor pressure is a measure of the evaporation rate, which is equal to the condensation rate at equilibrium. Yet the saturation vapor pressure also depends in a roundabout way on the condensation rate. More precisely, it depends on the relation between the condensation rate and the number density of condensing molecules. The simplest way to explain this is to return to the leaky bucket analogy of Chapter 2.

Flow into and out of the bucket are analogous to evaporation and condensation, respectively; the equilibrium water level is analogous to the saturation vapor pressure. If the inflow is increased, so is this water level, provided that the hole in the bucket is fixed. But if it were made smaller, the inflow remaining constant, the water level would rise until the outflow once again equaled the inflow. So the equilibrium water level depends not only on the inflow but also on the size of the outlet hole. The smaller this hole, the higher the water level must be before inflow and outflow are equal.

The proportionality factor between the rate of condensation and the number density of vapor molecules is analogous to the size of the hole in the bucket. To good approximation, this factor is independent of the presence of air mixed with the vapor. But as the amount of air is increased, all else being equal, the rate of condensation should decrease slightly. Thus the vapor density (hence the saturation vapor pressure) must increase slightly so that condensation again balances evaporation.

The net effect from all causes of increasing the air pressure is to *increase* the saturation vapor pressure. It is a very small change (e.g., increasing the air pressure by one atmosphere increases the vapor pressure by a fraction of a percent), but it is not zero. I have adduced this example merely to refute the notion that as the separation between its molecules increases air can hold more water vapor.

A PARTING SHOT

No explanation stands alone: it has corollaries. If the corollaries are silly, so is the explanation, no matter how well it serves limited purposes. Good explanations are crafted in the same way that a good pool player aims not only to sink a particular ball but also to position the cue ball so that after each shot he can sink yet another ball.

Rising-air-expands-cools-and-cold-air-can't-hold-as-much-water-as-hot-air-so-a-cloud-forms is an explanation of cloud formation to be expelled in one breath. But its corollaries are absurd, as I have tried to show by giving merely a few examples.

Lest you think that I am quibbling over words (but what is an explanation except a string of words?), let me give an example from my own experiences. I have asked the following question on examinations: "Wet wash is often hung out to dry even when temperatures are well below freezing. Is this just a waste of time?" I have received the following answer: "It is a waste of time. The cold air can't hold the water." Here is an example of the power of a slogan, frequently repeated, to paralyze rational thought. Under its spell, the author of this answer suspended common sense. I can remember my grandmother hauling great piles of damp clothes out onto the line in the depths of winter, and several hours later bringing in the slabs of frozen long underwear. I assume that she did this because she wanted to dry the clothes, not because she was atoning for her sins by backbreaking, but pointless, labor.

I could give more examples to show the poverty of the windbag argument. You can amuse yourself by devising others. But I am afraid that Duncan Blanchard is probably right. The windbag argument has transcended science. It has been made so many times that it has become true by repetition. It is an incantation, an article of faith, an unthinking act like the turning of a prayer wheel. It is a slogan, one with about as much intellectual content as those used to sell soaps and cigarettes—and it is likely to endure forever.

5 Mixing Clouds

I will have nought to do with a man who can blow hot and cold with the same breath.
Aesop: *The Man and the Satyr*

One morning in early spring, while slumped in a comfortable chair and staring vacantly into the kitchen, I noticed something that aroused me from my torpor. I had just returned from a brisk walk with our dog. He was lying on the kitchen floor panting heavily while my wife brushed him. A small cloud formed momentarily on every breath he exhaled. Yet my wife's breath was invisible. So I asked her to pant harder. Still no cloud. I begged her to try harder. But despite her best efforts—after more than twenty years with me she is unfazed by my often strange requests—no clouds came forth on her breath. This got me to thinking about mixing clouds.

Mixing clouds, one example of which is the clouds sometimes formed on your breath when it is cold, do not yield precipitation. But this does not mean that they are meteorologically insignificant. To me they are one of the most interesting kinds of clouds because they occur in so many different places and on a smaller scale than the clouds that yield rain and snow. To explain why mixing clouds form I shall first appeal to a simple demonstration. This will lay the groundwork for resolving the difference between my dog's breath and my wife's breath.

A MIXING CLOUD DEMONSTRATION

The idea for this demonstration comes from R. A. R. Tricker's *The Science of Clouds,* a delightful little book which, unfortunately, is out of print. All that you need is a vessel with a nozzle, a hot plate, and a small gas torch. If you heat some water in the vessel—I have used a vacuum flask—it won't be long before a cloud like that shown in Figure 5.1 issues from the nozzle just as it does from the spout of a whistling tea kettle on the boil. Such a cloud is popular-

Figure 5.1 A cloud forms when the hot water vapor issuing from the nozzle mixes with the colder surrounding air.

ly, although incorrectly, called "steam." But steam is merely water vapor at a temperature equal to or greater than the boiling point. We cannot see steam, so what we can see is not rightly called steam. Call the cloud what you like, just remember that it is composed of water droplets not water vapor. It will become apparent that a more descriptive term for this cloud is a *mixing cloud* because it is the mixing of moist air masses with different characteristics that yields such a cloud.

If you now heat the nozzle with a gas torch the cloud will soon disappear, as shown in Figure 5.2. Yet water vapor is still issuing from the nozzle. Why does a cloud form in one instance but not another?

For our purposes a given quantity of moist air—call it a parcel for convenience—may be characterized by two quantities: its temperature and the partial pressure of the water vapor in it. Suppose that two parcels with different temperatures and vapor pressures mix somehow. It is reasonable to expect the temperature and vapor pressure of the resulting mixture to lie somewhere between those of the two parcels from which it was compounded. What these intermediate values will be depends on the ratio of masses of the two parcels. For example, if a large mass of warm air mixes with a much smaller mass of cold air, the temperature of the mixture will lie closer to that of the warm air than to that of the cold air. This is hardly startling. What may come as a surprise is that the two parcels may be parents to offspring with qualitatively different characteristics, and this is the key to understanding why clouds may or may not form when moist air masses mix. To clarify this, a simple diagram helps.

The state of a parcel may be represented by a point in a plane, one coordinate of which is the temperature, the other coordinate of which is the vapor pressure. At any given temperature there is a special pressure (see Chapter 2), called the

Figure 5.2 By heating the nozzle with a gas torch a mixing cloud does not form although hot water vapor still issues from the nozzle.

equilibrium or saturation vapor pressure, which is the partial pressure of water vapor in equilibrium with liquid or solid water. Since the equilibrium vapor pressure depends only on temperature, equilibrium states lie on a curve, called the equilibrium curve, which is shown in Figure 5.3 on page 32. If the point representing a parcel's state lies below the equilibrium curve, the air is said to be unsaturated. If it lies above the equilibrium curve, the air is said to be supersaturated. And if it lies on the curve, the air is said to be saturated (see Chapters 2 and 4 for further discussion of the concept of equilibrium vapor pressure and a caution against misinterpreting the terms saturation and saturated).

Two parcels that are themselves unsaturated may mix to form a supersaturated parcel: the offspring has a characteristic not possessed by its parents. And if condensation nuclei are present—they almost always are—a cloud forms. An example of this is provided by the first part of the demonstration.

Mostly water vapor at a temperature of 100°C (212°F) and at atmospheric pressure emerges from the nozzle of the flask shown in Figure 5.1. We may represent the state of this water vapor by the point N in Figure 5.3; this saturated water vapor mixes with the much cooler and drier surrounding air, the state of which is represented by the point S. The states of all possible parcels that result when the saturated water vapor mixes with the surrounding unsaturated air lie (approximately) on a straight line joining N and S. Where precisely the state lies on this line depends on the relative masses of the two parcels that go into the mixture. What is important is that part of the line lies *above* the equilibrium curve. In this instance, therefore, a mixing cloud may form.

Note that whether or not such a cloud may form depends very much on where the two points N and S representing the two parcels lie. We can certainly imagine pairs of points such that the line joining them lies below the equilibrium curve. Indeed, this helps to explain why we could suppress cloud

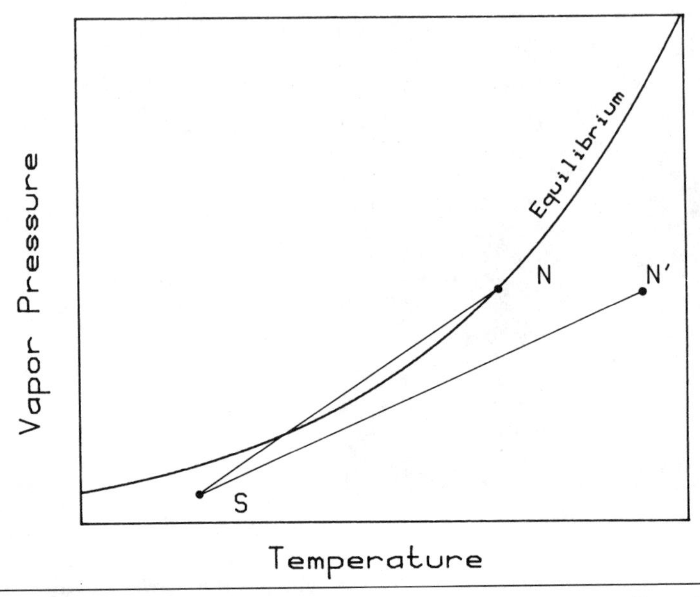

Figure 5.3 Why mixing clouds can form when moist parcels mix. Part of the line joining points N (nozzle) and S (surroundings) lies above the equilibrium curve. In this instance, two unsaturated parcels can mix to form a supersaturated parcel. But if N moves to N′ no mixing cloud is possible because the line joining N′ and S lies below the equilibrium curve.

formation by heating the nozzle. The emerging water vapor is still at atmospheric pressure, but its temperature is greatly increased, that is, the point representing its state moves to the right of N to N′. The straight line connecting N′ and S lies below the equilibrium curve. In this instance, therefore, a mixing cloud cannot form.

I must emphasize that the figure discussed in the preceding paragraphs is only schematic, it is not drawn to scale. This is why I purposely omitted labeling the units of temperature and vapor pressure. Equilibrium vapor pressure increases so much in going from room temperature to the boiling point and above that it would have been difficult for me to explain the demonstration with an accurately drawn figure.

It is sometimes stated that mixing clouds form because warm, moist air is cooled when it mixes with cold air. But it would make just as much sense to say that the cold air is heated. It makes even more sense to invoke neither heating nor cooling. What is essential to the formation of mixing clouds is the *shape* of the equilibrium curve. If it were a straight line, for example, clouds would never form when two unsaturated parcels of air mix regardless of how different they were. It is only because the equilibrium vapor pressure increases so rapidly with increasing temperature that mixing clouds form readily.

LAMINAR AND TURBULENT FLOW

Now that we understand the essentials of cloud formation by mixing, let us consider some of the finer points of the demonstration such as the wedge shape of the cloud and the gap between the nozzle and the cloud.

Water vapor flows smoothly from the nozzle at first, but a short distance from it the flow abruptly becomes chaotic. This is an example of a transition from laminar to turbulent flow. Laminar flow is smooth, regular; turbulent flow is chaotic, irregular. The distinction between these two types of flow is evident in a stream of water flowing from an ordinary kitchen faucet (without an aerator). Water flowing slowly from the faucet almost looks like clear glass; you can see objects behind the stream. But increase the flow a bit and suddenly the clear stream appears turbid. Another example of a transition from laminar to turbulent flow is provided by a lit cigarette on the edge of a table in a still room. Smoke rises in a regular stream from the tip of the cigarette, but at a short distance from it this smooth stream suddenly becomes chaotic. The smoke has almost nothing to do with the change in characteristics of the flow, it is merely a tracer. A hot object the same size and temperature as the cigarette tip would have the same pattern of air flowing above it, but this would be invisible without some particles in the flow.

Turbulent flow is characterized by rapid and violent mixing. When hot water vapor emerges from the nozzle it does so smoothly. It therefore does not mix much with the surrounding cooler air. No mixing, no cloud. But when the flow becomes turbulent a short distance from the nozzle there is rapid mixing and a cloud forms readily. The gap near the nozzle is therefore the region over which the flow is laminar; beyond the gap it is turbulent.

Diffusion is the random transport of matter. It may be demonstrated by carefully pouring water into a vessel in which there is water colored with a dye of some kind. With luck—and a steady hand—the boundary between the colored and uncolored water will be distinct. But with time, this boundary will become less distinct. After a long time—many hours, even days—the boundary will have disappeared: all the water will be colored. By random motions individual dye molecules have migrated across the boundary. Initially, there were many more of them on one side than on the other. Thus there was a net migration of dye molecules into the uncolored water. But after a time the concentration of dye increased to the point where its molecules cross the boundary at the same rate in both directions. This is *molecular* diffusion, which is exceedingly slow. *Turbulent* diffusion is much faster; it may be looked upon as the migration of globs of fluid called eddies. In molecular diffusion molecules migrate individually; in turbulent diffusion groups of many of them migrate together.

Where the jet emerging from the nozzle is laminar, water vapor diffuses slowly, molecule by molecule, into the surrounding air. But in the region of turbulent flow there is very rapid turbulent diffusion of water vapor in directions perpendicular to the jet. This is why the cloud has a wedge shape. The farther from the nozzle the more time there has been for water vapor to diffuse turbulently into the air, and a greater amount of time for diffusion corresponds to a greater distance over which it occurs.

ONE DIFFERENCE BETWEEN DOG AND WIFE

I began this chapter with a problem. The time has come to solve it.

Recall that one morning in the kitchen my dog's breath yielded mixing clouds but my wife's breath did not, no matter how hard she tried. It was a cool morning, and we keep the temperature of our home at about 15°C (60°F); the relative humidity was probably low. The state of the air in the kitchen is represented by the point K in Figure 5.4.

The deep body temperature of dogs is somewhat higher than that of humans, perhaps a few degrees. Of course, what determines whether mixing clouds are formed is not so much deep body temperature as the temperature of exhaled air. When you want to produce mixing clouds on your breath, or warm your hands with it, you do so with warm air from your lungs not with cooler air from your mouth. But even air from your lungs is likely to have cooled somewhat before it is exhaled. If we take the deep body temperature of humans to be 37°C (98.6°F), then a good guess for the maximum temperature of exhaled air might

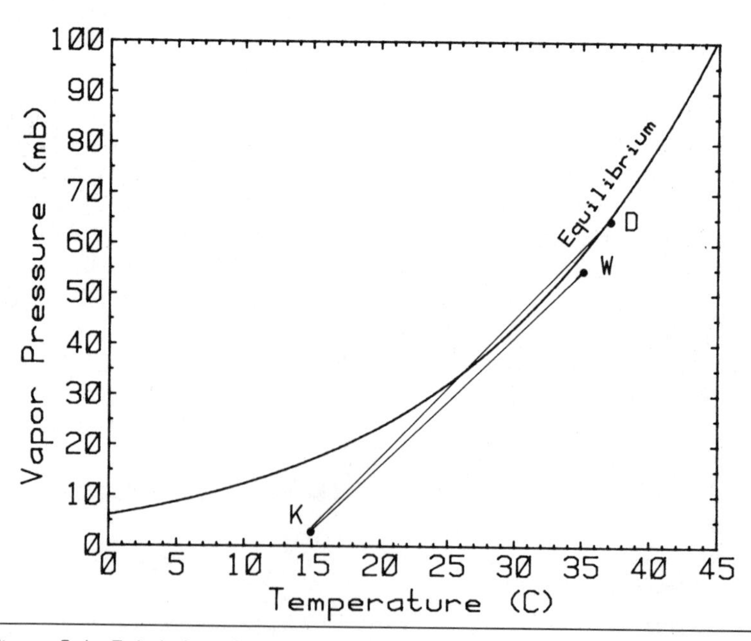

Figure 5.4 Exhaled air from a dog (D) is hotter and more humid than that from a woman (W). Consequently, the dog exhales a mixing cloud whereas the woman does not.

be 35°C (I have measured mouth temperatures with an infrared thermometer, and they are usually 32–34°C). The temperature of my dog's breath was probably a few degrees higher, say 37°C or more, especially after a morning run. It is also likely that his breath was more humid than my wife's. After all, he depends on panting for cooling. The points W and D represent the states of my wife's and dog's breaths, respectively. Part of the straight line connecting D and K lies above the equilibrium curve. But the line from W to K lies entirely below it. Thus because of his slightly warmer and more humid breath my dog succeeded where my wife failed (she has compensating qualities, however).

I hope that it is now clearer why clouds form on your breath on some days but not on others. For a fixed temperature and humidity of exhaled air, mixing clouds are more likely the colder and more humid the surrounding air. This is why you see clouds on your breath often in winter but rarely in summer.

CONTRAILS AND OTHER MIXING CLOUDS

Although it is difficult to fly *to* State College, Pennsylvania, where I live, it is almost impossible not to fly *over* it. On any clear day the sky is streaked with contrails like that shown in Figure 5.5. Contrails—from *con*densation *trails*—are yet another example of mixing clouds.

A byproduct of the burning of hydrocarbon fuels in jet engines is water. When this hot water vapor from the engine's exhaust mixes with the surrounding air,

Figure 5.5 When hot water formed by combustion in the engine of a jet aircraft mixes with the colder surrounding air this contrail results. It owes its existence to processes different from those that gave rise to the clouds below it.

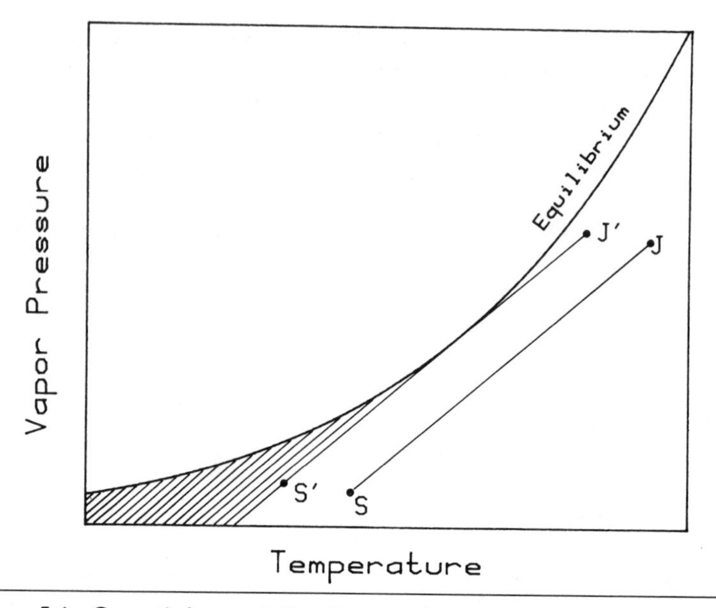

Figure 5.6 Contrails form only if ambient conditions are favorable. The hatched region shows all those states of air in which a jet aircraft would produce contrails. This equilibrium curve is only schematic because of the huge difference between the temperature of the ambient air (usually well below freezing) and that of the jet exhaust.

a cloud in the form of a contrail may result. Whether it does or not depends on the state of the surroundings.

Ambient air enters the jet engine. When it leaves it has been heated and has had water vapor added to it. Moreover, the temperature and vapor pressure rise are proportional to one another because they arise from the same cause: combustion. Consider the schematic diagram in Figure 5.6. Let S represent the state of the air through which a jet aircraft is flying. The state of the exhaust gas is represented by the point J, and all subsequent states of exhaust air mixed with surrounding air lie on the straight line connecting J with S. In this instance, no contrail forms because this line lies below the equilibrium curve.

Suppose now that the ambient air, represented by S′, is both colder and more humid. The exhaust air J′ is correspondingly colder and more humid; that is, the line through S′ and J′ is parallel to that through S and J. If J′S′ is tangent to the equilibrium curve then the latter is the boundary that separates air in which contrails can form from that in which they cannot. The hatched region of the diagram shows all those states of the surrounding air for which a contrail can form in a jet's exhaust. This region is determined by the characteristics of the engine, which determine the slope of the boundary line, and the equilibrium curve.

Once you begin looking for them, mixing clouds can be found almost anywhere: in the plumes of power plants; in the exhausts of automobiles (why are they so thick immediately after starting but then gradually disappear?); in air rising from a cup of hot coffee or soup; above ponds in winter. You can amuse yourself by adding to this list.

6

Conceptions and Misconceptions of Pressure

"I'm becoming frightened, Doctor," she said, in a low voice. She smiled as she spoke, but I could see that she was seriously worried. "Do you know what the barometer reads? Twenty-eight seventy! And the wind is growing stronger every minute."

Charles Nordhoff and James Hall surely would not have included this scene in their rattling good tale, *The Hurricane,* had they doubted that a mere barometer reading would convey an apprehension of impending disaster. For who, however rudimentary his knowledge of meteorology, does not associate the rise and fall of barometric pressure with changes in the weather? Indeed, there is no physical quantity the variation of which in time and space is more important to weather forecasting. What, then, is pressure? Although it is true that sea level atmospheric pressure is a measure of the weight of the atmosphere, this is not sufficient to explain all phenomena associated with atmospheric pressure and its variations. Moreover, a fluid's pressure and its weight are two different concepts; they are not synonymous, for if they were then one of them could—and should—be discarded, at no loss to physical understanding and at a gain in economy.

Like temperature, pressure in its everyday sense is a prescientific concept, and you do not have to know what it is to experience it. This is obvious to anyone who dives under water and feels his surroundings squeeze him more tightly the deeper he descends. This experience of pressure is as old as humankind. Yet only comparatively recently did it become evident that the atmosphere, like water, is also a fluid capable of exerting a pressure. This realization was long in coming because we do not ordinarily sense atmospheric pressure. Awareness of atmospheric pressure comes with the ability to measure it, and instruments to do so were not invented until the seventeenth century.

DEMONSTRATING ATMOSPHERIC PRESSURE

There are so many demonstrations of atmospheric pressure that it would be futile to labor mightily on devising a new one or varying somehow an old one. Instead, I shall describe the one I consider to be the best of the lot. It was reported by Pieter Visscher of the University of Alabama in a short note to the *American Journal of Physics* (1979, Vol. 47, p. 1015).

Very little is needed for this demonstration: a hotplate, an empty soft (or hard) drink can, and a shallow pan filled with water. Put a small amount of water (about a spoonful) into the can and place it on the hotplate. The water

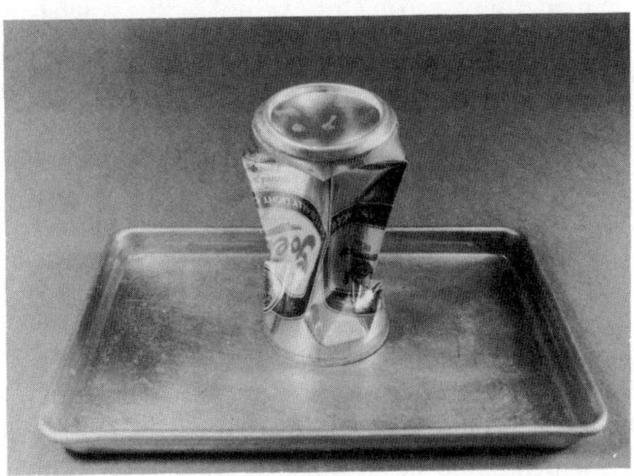

Figure 6.1 When a can filled with hot water vapor is placed in a shallow pan filled with cooler water, the can is crushed because of the pressure difference between its inside and its outside. That outside is atmospheric pressure. That inside is initially atmospheric pressure but drops as the water vapor condenses onto the cooled walls of the can.

temperature will rise quickly and the can will fill with water vapor. Unlike the gaseous constituents of the atmosphere, water vapor is readily condensed (this is the distinction between a gas and a vapor). To do so just snatch the can from the hotplate and plunge it mouth first into the pan. In a few seconds, if all goes well, the can will be crushed (see Fig. 6.1) accompanied by a dull pop. Very dramatic. It sometimes takes more than one try depending on how successful you are at finding something with which to grasp the can.

Tongs of the proper span would be best, but these never seem to be available so I often improvise them using whatever is at hand. As a consequence, I sometimes drop the can or it ends up on its side in the water—a fizzle. But never mind, the demonstration is easily repeated, and even the most fumbling efforts are usually met with success. Be sure to use a *shallow* pan. If you do not, all that is likely to happen is that the can will sink into the water without being crushed.

In Chapter 2 I discussed the concept of saturation vapor pressure—the pressure of a vapor in equilibrium with its liquid—and in Chapter 5 I emphasized how rapidly it increases with increasing temperature. For example, in going from 40° to 100°C (the normal boiling point) the saturation vapor pressure of water increases by more than a factor of ten. It is this large change that crushes the can.

The can sizzling on the hotplate was filled mostly with water vapor at a pressure about that of the surroundings (at the boiling point the saturation vapor pressure is that of the surroundings). When this hot can was plunged suddenly into room-temperature water it cooled, hence the rate of evaporation from its inner surface decreased. The rate of condensation onto it, however, was slower to respond. So there was *net* condensation of water vapor for a moment, and the pressure inside the can consequently dropped to well below that of the surroundings. Thin aluminum cans are not able to withstand such large pressure differences so they collapse. You may wonder why the pressure difference doesn't merely force water into the can, like mercury in a barometer. With a much stouter can—one made of cast iron, for example—this is what would happen. And it would also happen even with a thin aluminum can were it not for the inertia of the water: before the water has sufficient time to flow in, the can collapses.

This demonstration of the *existence* of atmospheric pressure is unambiguous. It is only when we *interpret* what we observe that the ambiguities and contradictions arise.

It is often asserted that atmospheric pressure is just the weight of the atmosphere above a unit area. While this is true in the absence of vertical acceleration, it does not come to grips with the concept of pressure in a fundamental way. For gas pressure and weight are, in general, independent. To convince yourself of this consider the following thought experiment. Fill a sealed container with a gas, which could be air but need not be. At a given location the weight of this gas—that is, the force exerted on it by the gravitational attraction of the earth—is fixed. Yet it is a matter of experience that the gas pressure would rise if the container were heated. And if the gas were taken beyond the pull of the earth's gravitational field it would become weightless but not pressureless. Thus the relationship between pressure and weight is not unique. Moreover, the weight interpretation of pressure does little to help us understand the demonstration. There is more to the pressure concept than that it is an alternative guise for weight.

INTERPRETING PRESSURE

To our coarse senses all matter is *continuous:* we can discern no gaps, no lumps. This is particularly true of the common fluids air and water. Despite this, the hypothesis that matter is ultimately *discrete,* the atomic (or molecular) hypothesis, has had many adherents almost since the beginning of recorded thought. And its appeal is not to be wondered at, for we too are atoms. This does not mean that we are very small: the root meaning of atom, from the Greek *atomos,* is indivisible. Neither we nor atoms can be divided, at least not without losing our salient characteristics: cut a man in two and he just won't be the same.

Although the existence of atoms was hotly debated for centuries, it was only during the first decade of this century that enough evidence had been amassed to silence the critics, among whom were counted some of the most eminent scientists of the day. Today, anyone who seriously denies the existence of atoms and molecules (although not strictly indivisible, a molecule is a more or less stable aggregation of atoms) is classed with the flat-earth believers. But in one sense the critics of the molecular hypothesis were right: it is sometimes neither necessary nor even desirable to take account of the discreteness of matter. For many purposes it is more expedient to proceed *as if* matter were continuous.

Few people have not at least heard of molecules. Unfortunately, what they have heard is often so fragmentary, so superficial, so distorted, that it would have been better had they not. A little bit of knowledge is sometimes less than none at all. For example, if one adopts the continuum interpretation of matter, then one has little choice but to consider the pressure within a gas as resulting from adjacent portions of a continuous medium pushing against one another. If one now switches to the molecular interpretation it is only natural to carry over this notion, that is, to consider gas pressure as resulting from gas molecules pushing against one another. As attractive and widespread as this notion is, it is nevertheless false. The error arises in going from one interpretation to another. Translation errors, sometimes the cause of embarrassment, are not uncommon in going from one language to another. Concepts in one language do not necessarily have exact counterparts in another. So also is it with different languages for describing the physical world. Let us explore how one such language, that of atoms and molecules, describes gas pressure.

MOLECULAR INTERPRETATION OF GAS PRESSURE

Even after one adopts the molecular viewpoint, one is confronted with two aspects of the concept of gas pressure: as a *measurable* quantity and as a *property* of the gas. I shall consider each of these in turn.

Consider a gas in a rigid container. In even a tiny volume of this gas there is an enormous number of molecules: in one cubic millimeter, more than a million billion of them (under normal conditions). Molecules that collide with the walls of the container will suffer a change in *momentum*. Momentum, which is the product of mass and velocity, is a directed quantity. Thus even if the speed of a molecule—how fast, but not where, it is going—does not change upon collision, its momentum does: before collision it was moving *toward* the wall, after collision it is moving *away* from it.

Newton's second law states that a body will not change its momentum unless a force acts on it. And from his third law—to every action there is an equal and opposite reaction—it follows that if the wall exerts a force on a molecule that collides with it, each such molecule exerts an equal and opposite force on the wall (this is painfully obvious to anyone who has ever run headlong into a wall). Because of their great number and speed (at room temperature the average speed of an air molecule is about that of a rifle bullet) molecules collide with the wall at a tremendous rate: in a millionth of a second, each square centimeter of wall is struck by billions upon billions of molecules. And each of them contributes its mite to the pressure, the total force acting on a unit area of the wall. This is the force that causes the liquid in barometers to rise and that caused the can to collapse: the total force due to molecular collisions with the inside of the can was overwhelmed by that due to collisions with the outside.

Note that there is an important distinction between the target (i.e., a surface) and the projectiles (i.e., gas molecules) impinging on it. The average separation between molecules in a gas (at normal temperatures and pressures) is about ten times that in a solid or liquid. The density of solids and liquids is therefore about a thousand times greater than that of gases. As a consequence, the rate at which gas molecules collide with one another cannot be anywhere near the rate at which they collide with a liquid or solid surface. Indeed, to good approximation air may be treated as an *ideal* gas, another name for which is a *collisionless* gas. The very name gives the game away. It is not that collisions between gas molecules do not occur, or even that they are not important in determining *some* properties of gases, it is just that they are not the determinants of such quantities as gas pressure. As far as pressure is concerned, the molecules might just as well be collisionless—with one another, of course. And the same holds true for temperature. Two gases with greatly different rates of intermolecular collision (e.g., two gases with greatly different densities) can be at the same temperature.

When gas pressure is measured, it is always by virtue of molecules interacting with a solid or liquid surface (e.g., a barometer). If we were concerned only with the measurability of pressure then we would need say no more than this. But what meaning is to be attached to pressure construed as a property of the gas? That is, how should we interpret pressure within a gas as opposed to at its interface with a liquid or solid?

The dimensions of pressure are force per unit area, as are those of *momentum flux,* the rate at which momentum is transported across a unit area. This is not just a coincidence. Recall that pressure was associated with the rate at which gas molecules transfer momentum to unit area of a material surface at the *boun-*

dary of a gas. Now imagine a point *within* a gas. Molecules in the neighborhood of this point are whizzing about in all directions, but their total momentum is zero if the gas is at rest: for every molecule with a given momentum there is another with the opposite momentum. Contrary to what you might expect, however, the total momentum flux is not zero. Consider an arbitrary direction, call it the positive direction (the opposite direction is the negative direction). Each molecule has a component of momentum, either positive or negative, along this direction. Moving molecules carry their properties, one of which is momentum, with them wherever they go; that is, molecules transport momentum. The reason that the total momentum flux is not zero, even though the total momentum is, is that positive momentum transported into a region is equivalent to negative momentum transported out. A simple analogy comes to mind: getting rid of a debt (an outflow of negative money) is equivalent to a raise in salary (an inflow of positive money). Lest I mislead anyone, I must emphasize that I have in mind momentum transported across a unit area of an imaginary *flat* surface in a gas. The net rate at which momentum is transported across a *closed* surface in a gas is zero if there are no forces acting on the molecules.

According to the molecular interpretation, the pressure in a gas is the total rate at which its molecules transport their momentum, along a given direction, across a unit area perpendicular to that direction. Regardless of the direction chosen, the corresponding momentum flux is the same. Consequently, the pressure acts equally in all directions. Pressure has essentially nothing to do with the rate at which gas molecules collide with one another. This rate is small compared with the rate at which they collide with a solid or liquid surface, and it is by means of the latter that pressure is measured.

How does the molecular interpretation elucidate observations of atmospheric pressure, such as its decrease with height? This is usually explained by saying that the weight of the atmosphere above a given elevation decreases with increasing elevation. This is true enough, but the observation has an alternative explanation. At each successive elevation lower pressure implies lower momentum flux. This means that for any horizontal layer of atmosphere more vertical momentum goes in at its bottom than comes out at its top. What happened to this momentum? According to Newton's laws, momentum changes (e.g., decreases) only if caused to do so by a force. In this instance, the force is that of gravity. And weight is, after all, merely the force of gravity. So the two interpretations—microscopic and macroscopic—are in harmony. But they can be made to clash unnecessarily by mixing concepts appropriate to one with those appropriate to the other. Consistency is what is wanted. Once you have adopted a particular viewpoint, stick with it. If you must change, beware of translation errors.

The choice of interpretation is to some extent a matter of taste. More often it is guided by considerations of expediency. It would be foolish, for example, to try to do a molecule by molecule forecast of the weather. For this purpose the view that matter is continuous is the most sensible one. For other purposes, however, understanding is acquired only by recognizing the ultimate discreteness of matter.

7

Dew Drops on a Bathroom Mirror

Nothing . . . is absolutely true. The earth is not quite round. The sky is not quite blue. Rain isn't altogether wet.
Stephen Leacock

Several years ago, after class one day, a student brought me a small vial similar to a purse-size perfume atomizer; its label proclaimed that it contained a wondrous compound which when applied to eyeglasses would keep them from fogging. He wanted to know how it worked and whether it was worth its steep price. I told him that I would show him how to obtain a lifetime supply of antifogging compound for a few dollars.

When class next met I had with me two Petri dishes and a container of liquid dishwashing detergent. First I breathed hot and heavily on one dish, the way you do when you want to warm your hands on a cold day, and immediately put it on an overhead projector to show that it had been fogged by my breath. Then I smeared a very thin film of liquid detergent onto the other dish. A fraction of a tiny droplet is more than enough. This dish, with its imperceptible film of detergent, I could not fog with my breath, no matter how hard I tried. You might think that the film prevented water from condensing onto the dish—a natural, but incorrect, supposition. To convince yourself of this you can do a simple (and relaxing) experiment in your own home.

TAKE A HOT BATH

This experiment is done in the bathroom, a home laboratory second in importance only to the kitchen. With liquid soap or detergent make a figure on the dry bathroom mirror; the film need be only very thin. Then hop into the tub and take a good hot bath, preferably on a cold day. Take your time. Enjoy your soak. And as you blissfully stew in your own juices cast a glance now and again at the mirror.

The moist air in the bathroom is much warmer than the dry mirror. From the mirror's point of view, its environment is supersaturated (see Chapter 5). Water will therefore gradually condense onto it, as evidenced by fogging—except where it is covered by the soap film. To convince yourself that the spot which appears to be free of water is not, press small pieces of tissue paper against clear and fogged spots. Paper will stick to both of them (shown in Fig. 7.1), but not to the mirror when it is dry. So the soap film has not prevented water from condensing.

The fogged part of the mirror is covered with small water droplets, whereas the clear part is covered with a more or less continuous film of water. Scattering has not been prevented, only its nature has been changed: the water droplets scatter light in all directions; the water film scatters light mostly in a few directions (for more on this see Chapters 13 and 14).

There is a way to prevent (net) condensation, which I had to use in order to take the photograph in Figure 7.1. On my first attempt I got out of the tub, took my camera and tripod from a cold bedroom, set them up outside the bathroom, opened the door poised to shoot—and couldn't see a thing. My lens had fogged. Of course, I should have anticipated this, but perhaps my critical faculties had been softened by the warm waters of my bath.

To prevent the lens from fogging, I took off the haze filter and heated it by suspending it on a roasting rack over a lamp. Water vapor is still incident on

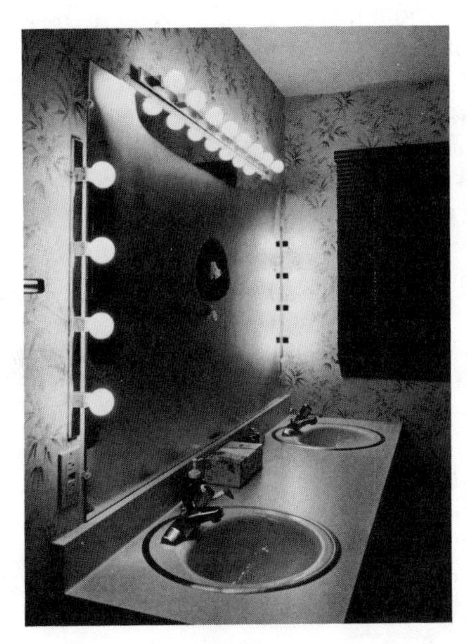

Figure 7.1 This mirror is fogged everywhere except in a spot over which a thin soap film had been smeared. That the film does not prevent water from condensing onto the mirror is evident from the two pieces of tissue paper sticking to the wet mirror, both where it is fogged and where it is clear. Paper will not stick to the mirror when it is dry.

the hot filter but it evaporates as fast as it condenses so there is no net condensation. This is also exemplified by the dry parts of the mirror warmed by the lights.

After applying the soap film to my bathroom mirror I took well over a dozen baths, and each time the mirror did not fog on the spot with the soap film, which attests to its robustness.

Despite the familiarity of the fogging of mirrors, to understand it requires a critical and careful examination of some concepts which are often treated superficially. Among these is surface tension, which to invoke is not necessarily to understand. I shall therefore devote three sections to surface tension: what it means, whether it is real, how to demonstrate it. In so doing I shall point out that the surfaces of liquids are like miniature atmospheres, and then show you how to convince yourself of this similarity. A discussion of what happens to surface tension when soap is added to water leads to the more general concept of interfacial tension. Only then shall we be ready to go back to the fogging of mirrors, and finally make a perhaps unsuspected connection between it and nucleation of cloud droplets in the atmosphere.

SURFACE TENSION: FACT OR FICTION?

You have probably seen, and perhaps envied, long-legged bugs, called water striders, gliding on the surfaces of lakes or streams. They do not sink into the water because of surface tension, which is often explained by saying that it is "as if" there were an elastic skin, like that of a balloon, on the surface of the water. The water strider is so light that it is supported by surface tension just as we can poke the surface of a balloon without breaking it. Yet I can cite eminent authorities who, with just a hint of disdain, assert that the elastic skin on water is merely a convenient fiction, not something real. It is an intellectual crutch, no more needed by real scientists than riders in the Tour de France need training wheels. Yet there are other authorities, equally eminent, who assert that no one need be ashamed of invoking tensions at the surfaces of liquids because they are just as real as anything else in this shadowy world we inhabit. As so often happens in courts of law, the experts cancel each other's testimony and the jury is forced to do its own thinking.

Those who scoff at surface tension offer in its place surface energy. It is indeed true that it takes energy to form a surface, which is well known to anyone who whittles. This energy can be said to reside at surfaces, and we can consider the total energy of any object to be the sum of energies proportional to its volume and to its surface area.

Consider a small water droplet. To change its volume takes energy. But so does changing its shape. A sphere has the least surface area for given volume. To distort it we must therefore increase its surface area, hence its surface energy.

The spherical shape of small water droplets can be explained in two ways. Because a sphere has the least surface area a water droplet will assume this shape since all systems evolve in such a way as to minimize their energies. The other explanation is that surface tension forces the water droplet into a spherical shape. Which of these explanations is correct? What is really happening?

We could debate forever on what is really meant by "real." But instead of wrestling endlessly with philosophical conundrums we can merely try to decide if those who plump for the reality of surface tension are fooling themselves or not. Explaining observations by invoking the tendency of systems to evolve to states of lowest energy has its appeal, although we are still left wondering why this tendency exists. It might be more satisfying if we could imagine, without deluding ourselves with a faulty analogy, that real forces drive systems to their final states. We experience forces directly as muscular pushes and pulls, hence they are less abstract than energies. Yet it is difficult to accept that liquids are in tension at their surfaces. This is certainly not true in their interiors, and for this reason the proponents of surface energy may have a point. We are caught between two opposing camps. Surface energy is more abstract than surface tension, whereas the reality of the latter seems difficult to accept.

THE REALITY OF SURFACE TENSION

We are taught about the three phases of matter: gaseous, liquid, solid. Like all classifications this one is not absolute. Above the surface of water, matter is in the gaseous phase. Within the water, it is in the liquid phase. But in a thin region near the surface, matter is in neither the one nor the other phase, rather it is in what we have every right to call the surface phase. Having accepted the existence of this new phase, we should not be surprised if its properties are quite unlike those of the phases between which it lies.

A molecule moving about within a liquid is now attracted now repelled by surrounding molecules but on average experiences no *net* force because its environment is about the same in all directions. A molecule near the surface, however, experiences a net attractive force inward: there are fewer molecules above the surface to attract it outward. One way to describe this is to say that the surface is a region of higher potential energy than the bulk. Potential energy—energy of position, as opposed to kinetic energy, energy of motion—is the amount of work necessary to move an object from one position to another. For example, the gravitational potential energy of any object increases with height above the earth. And as anyone who has ever dropped a bag of groceries knows, objects congregate where the potential energy is least.

For example, the number density of air molecules in the atmosphere is greatest where the potential energy is least, and conversely. That is, the pressure drops as we ascend mountains. Similarly, the density of molecules near a liquid surface is less than in the bulk of the liquid because the surface is a region of higher potential energy. Molecules on their haphazard journeys enter this region, of course, just as molecules can be found in the upper reaches of the atmosphere, but they are not encouraged to stay.

What are the consequences of this reduced molecular density near the surface? Within a gas, pressure as a property is interpreted as molecular momentum transfer. Pressure is measured when gas molecules collide with a boundary surface—the walls of a container, for example—thereby transferring momentum to it (for more on the molecular interpretation of gas pressure see the previous chapter). In gases (strictly speaking, ideal gases) the molecules do not interact because of the large distances between them. This is no longer true in liquids where the intermolecular separation is about a tenth that in gases. In liquids the molecules exert appreciable forces on one another, and these forces contribute to the *total* pressure. This total pressure, which is what we usually measure, is the sum of a *kinetic* pressure, associated with the motion of molecules, and a *dynamic* pressure (also called internal pressure, intrinsic pressure, and cohesive pressure), associated with the forces between molecules. These two terms come from the Greek *kinetikos,* of motion, and *dynamikos,* related to physical force or energy.

The kinetic pressure is proportional to the number density of molecules. Densities of liquids are about a thousand times greater than that of the normal atmosphere. Hence the kinetic pressure in a liquid such as water is about 1000 atmospheres. Yet when we dive under water in a swimming pool we certainly do not experience such bone-crushing pressures. This is because the dynamic pressure nearly balances the kinetic pressure. Both pressures are large, about equal in magnitude, but opposite in direction, kinetic pressure being a compression, dynamic pressure being a tension. The difference between the two is about one atmosphere.

Near the surface, where the density of molecules is lower than in the bulk, both the kinetic and dynamic pressures are less. But the dynamic pressure decreases less than the kinetic pressure because increasing slightly the separation between molecules that are very close together can actually increase the attractive force between them. Near the surface, therefore, the tensile dynamic pressure is greater than the compressive kinetic pressure, and the resultant is a tension: surface tension.

DEMONSTRATIONS OF SURFACE TENSION

Now that I have convinced you, I hope, that you may, without looking shamefully over your shoulder, consider the surface of a liquid to be in tension, I should like to discuss a demonstration of it which lends itself nicely to illustrating several points.

You have probably used a straw or a similar tube to transfer a liquid from one container to another. You insert the straw in the liquid, water say, then seal the end with your finger. When you withdraw the straw, a column of water remains in it. Before you withdrew it the pressure above the water column was equal to that of the surrounding atmosphere. As you withdraw the straw some of the water leaks out, although you may not have noticed this. Since the space above the water in the straw is closed, the pressure in it drops as its volume increases. Now the pressure inside the straw is less than that outside, and the pressure difference is sufficient to support a column of water. When you remove your finger, the pressures inside and outside equalize and the water flows from the straw.

I must emphasize that capillarity—the rise of liquids in narrow tubes—has almost nothing to do with why the water column remains in the straw. You can convince yourself of this by noting that water does not rise appreciably into the straw when it is inserted into water. Capillarity is indeed a surface tension phenomenon. The term itself comes from the Latin *capillaris,* which means of hair. Consistent with this origin, capillarity is appreciable only in hairlike tubes.

Have you ever tried to hold water in straws or tubes of different sizes? If you do you will discover that there is a diameter above which water will not stay in a tube no matter how carefully you seal its end. The existence of a critical tube diameter is a consequence of surface tension. When water is held in the tube, a small, somewhat flattened hemispherical drop is suspended from the end. For this drop to exist its weight must be balanced by the force of surface tension. Weight increases as the cube of the tube diameter, whereas the upward-directed surface tension force increases as the diameter. Hence there is some diameter above which surface tension can no longer balance the weight of the droplet.

This critical diameter is about 1 cm, although I have not determined it exactly because I don't have a sufficiently graded collection of tubes. All that I can say with certainty is that a tube of diameter 1.25 cm will support a water column whereas a tube of diameter 1.5 cm will not.

With a tube of diameter about 1.25 cm you can test the analogy between the surface of a liquid and the atmosphere. The greater the altitude in earth's atmosphere the lower the concentration of air molecules, which congregate in regions of lowest potential energy. Similarly, the surface of a liquid is relatively depleted of molecules because it is a region of high potential energy. Thus the surface region of a liquid is a miniature atmosphere. Let us consider the extent to which this analogy holds.

The scale height of the atmosphere (strictly, the troposphere, the lowest 15–20 km) is about 8.4 km; this is the distance over which the molecular density decreases by a factor of about one-third. If the average temperature of the atmosphere were higher the scale height would be correspondingly greater. In the limit of indefinitely high temperature (all else being equal), the scale height would be infinite and the molecular density would be uniform. We can interpret this by saying that as the temperature increases the molecular motion becomes more violent and stirs the atmosphere to a more uniform density.

By analogy, we expect surface tension to decrease with increasing temperature. Surface tension owes its existence to a nonuniform density of molecules, and as the temperature is increased the density should become more uniform. To verify this expectation I used a glass tube with an internal diameter of about 1.25 cm. If I was very careful, I could just barely retain a column of room-temperature water in this tube. But when I tried to do so with water near the boiling temperature, I almost always failed. The critical diameter is only weakly dependent on surface tension (it increases as the square root), which itself does not vary greatly (about 15 percent) in going from room temperature to the boiling temperature. Nevertheless, this experiment does provide evidence that surface tension decreases with increasing temperature, in accordance with what we expect on the basis of the molecular interpretation of surface tension.

Up to this point I have considered only the surface tension of pure water. What happens when we add something to it? To answer this, you can again use a straw or a glass tube. One with a diameter of about 1 cm will not support a column of water to which soap has been added, no matter how hard you try. Soap is a *surface active* agent: it appreciably decreases the surface tension. By random motion soap molecules find their way to the surface where they are encouraged to stay by the force fields of the water molecules. Thus the concentration of molecules near the surface increases and the surface tension decreases.

INTERFACIAL TENSION AND THE WETTING OF SURFACES

Closely related to the concept of surface tension (or energy) is that of interfacial tension (or energy): the tension at a solid-liquid or liquid-liquid interface, for example. Indeed, surface tension, without qualification, usually means the interfacial tension at an interface between a gas and a liquid.

To show the consequences of different interfacial tensions I prepared the surfaces of three glass microscope slides differently. On one I smeared a thin oil film; on another a thin soap film; I cleaned the third slide with concentrated sulphuric acid. Then I put water drops on each of them. Drops on the oil surface never lost their identity and were nearly hemispherical, but on the clean glass they spread into thin puddles. Drops on the soap surface quickly lost their identity: they spread and merged into a more or less continuous film.

A variation on this experiment is to put soapy and pure water drops on an oil-covered surface. The soapy drops will spread and cover the surface, whereas the pure drops will not.

To describe these experiments it is useful to introduce the *contact angle,* which is shown for three surfaces in Figure 7.2. This is the angle a liquid drop makes with the surface on which it rests. When the contact angle is zero, the surface is said to be perfectly wettable by the liquid. The other extreme, a contact angle of 180 degrees, is obtained by a perfectly nonwettable surface. A drop on such a surface touches it over a small area. Perfectly wettable and nonwettable surfaces are two extremes; surfaces are wettable to varying degrees between them.

The contact angle depends on the tensions at the various interfaces: liquid-air, solid-air, and liquid-solid. These, in turn, are related to the relative attractions between the different molecules. A surface is more wettable the greater the attraction its molecules have for those in the liquid relative to the attrac-

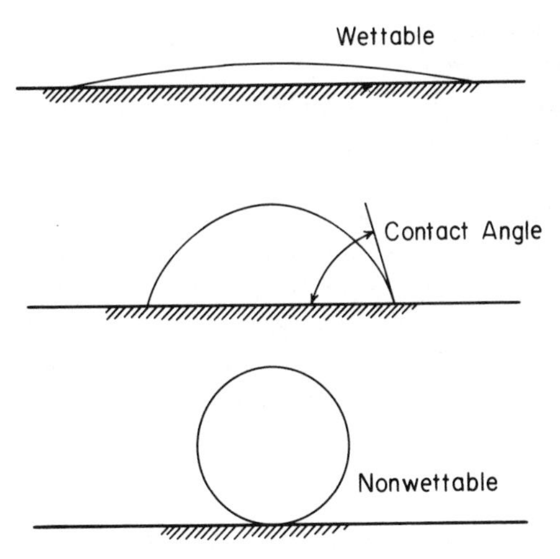

Figure 7.2 Water (or any liquid) wets various surfaces to different degrees, as specified by the contact angle. It is zero for a perfectly wettable surface and 180 degrees for a perfectly nonwettable surface.

tion those in the liquid have for each other. Molecules in the nonwetting droplet much prefer each other's company, and demonstrate this by curling up into a ball so that they are as far from the surface as possible. But those in the wetting liquid find surface molecules to be as agreeable as their own kind; a drop therefore spreads as freely over the surface as it would over its parent liquid.

You will find statements that the contact angle for water on clean glass is zero. Perhaps it is, but I wonder if those who make this unequivocal statement have ever tried to scrub glass so clean that its contact angle is zero. I failed to do so. The contact angle was small, but not zero. The lower reaches of Hell must be populated by sinners who, to keep them busy for eternity, have been given the task of preparing perfectly clean water and perfectly clean surfaces.

BACK TO THE BATHROOM MIRROR

Clean as my bathroom mirror might appear to the casual observer, it is quite filthy as evidenced by its contact angle with water. Droplets form on it nicely, which is shown in Figure 7.1. This mirror undoubtedly is contaminated, and to make it perfectly clean would be a task more Herculean than cleaning the Augean stables. But a thin coating of detergent reduces the contact angle to such a small value that water condenses onto the mirror as a thin film rather than as independent droplets.

Although mirror fogging is usually a nuisance, it can be put to good use, to determine the *dew point,* for example. This is the temperature to which air must be cooled, at constant pressure, in order for saturation to occur.

A mirror the temperature of which can be controlled is the heart of an automatic dew-point hygrometer. Light illuminates the mirror, oriented so that light reflected by it when it is dry will not be received by a detector. As the mirror is cooled to the dew point, the detector signal will increase suddenly as water droplets condensed onto the surface scatter light into the detector.

Is it possible that mystery writers have missed a plot? Invisible writing often figures in cloak-and-dagger stories. Bathroom mirrors could transmit secret messages only to those who know how to make them visible: take a hot bath. Here is the raw material for a thriller.

A friend of mine, Jorge Pena, once told me about a folk remedy for broken windshield wipers: keep a potato in the trunk of your car. If the wipers fail during a rainstorm, and you are far from help, smear your windshield with a fresh slice of potato. This is supposed to make the drops spread more into a film. I haven't put this method to a direct test, but I did rub a slice of potato over a spot on my mirror. And sure enough, it did not fog the next time I took a bath.

NUCLEATION IN THE ATMOSPHERE

In Chapters 2 and 4 I discussed nucleation of cloud droplets in the atmosphere. In particular, I pointed out that tiny droplets of water, if pure, cannot exist in equilibrium with water vapor except at relative humidities vastly higher than those in the atmosphere. But if such droplets form on soluble particles, they can exist at modest humidities because dissolution lowers the saturation vapor pressure. Yet there is no law that requires all particles in the atmosphere to be soluble. Suppose that none of them were? Would cloud droplets still form on them under conditions we consider to be normal?

It depends on the contact angle of the particles. Imagine that the three surfaces in Figure 7.2 are parts of different small particles. Can the water droplets on these surfaces be in equilibrium in the same environment? Of course not: the radius of each droplet is different, and saturation vapor pressure (that is, evaporation rate) depends on radius. The droplet on the nonwettable surface has the smallest radius of curvature, that on the wettable surface has the greatest. So if the environmental humidity were such that the droplet on the wettable surface neither grows nor shrinks, the other two droplets would eventually evaporate. This implies that a particle will be a better condensation nucleus (that is, accrete liquid water at a lower degree of supersaturation) the lower its contact angle. The theory of this was put forward many years ago (1939) by Max Volmer, and verified experimentally in 1959 by Sean Twomey.

Twomey's experiment was a refined version of what I did in my bathroom. He exposed surfaces with different contact angles to varying degrees of supersaturation and noted when they fogged. The greater the contact angle, the greater the degree of supersaturation required for fogging, in agreement with Volmer's theory.

And now we have come full circle. I began by pointing out that a soap film on a mirror does not prevent water from condensing onto it. Now I can go even further and say that the soap film actually makes it easier for condensation to occur.

8

A Murder in Ceylon

I went down to the St. James Infirmary
I saw my baby there
She was stretched out on a long, white table
So cold, so pale, and fair

Judged by his superbly written autobiography, *Mostly Murder,* the late Sir Sydney Smith must have been a charming and delightful man. But the field in which he achieved eminence, forensic pathology, could hardly be so described; morbidly fascinating might be more apt. The function of forensic pathologists is to determine how the dead met their end: murder, suicide, or accident. Sir Sydney chose to recount mostly the murders he investigated during his long, distinguished career, and one of them gives the title to this chapter.

About thirty years ago Sir Sydney was involved in a case of murder in Ceylon (now Sri Lanka). A man was accused of murdering his wife. His whereabouts were known reliably *after* a certain time, and the victim was known to have been alive *before* a certain time. Whether or not the accused was to be hanged therefore crucially depended on determining the time of death as accurately as possible. After death a body cools slowly at a rate depending on its size and clothing and its surroundings (still or windy, for example). From a body's *cooling curve* the time of death can be estimated by extrapolating backwards from when the temperature is known to when it may be assumed to have been normal (about 37°C). In this chapter I shall tell you how to construct cooling curves—not for corpses but for flasks of water—and what can be learned from them. As is done in all good detective novels, I save the resolution of the murder for the end.

NEWTON'S LAW OF COOLING

Almost three hundred years ago Sir Isaac Newton noted that a hot object cools in such a way that the difference between its temperature and that of its surroundings decreases *exponentially*. That is, if this temperature difference falls to half its initial value in a certain time interval, then in twice this interval the difference will be one-fourth the initial value, in three such time intervals it will be one-eighth, and so on. Strictly speaking, therefore, one must wait forever for the body to cool to the temperature of its surroundings, although in a finite time the temperature difference will be immeasurably small.

In Newton's day the distinction between temperature and heat was poorly understood. Even today the term heat is often used in contradictory ways: that an object "contains" so much heat, for example. This notion can be demolished readily enough with a simple demonstration. Take a small bottle and put some water in it, 10–30 ml, say. Measure the temperature of the water. Then cap the bottle and shake it vigorously for a minute or two. And by vigorously I mean that you should be panting and sweating from your efforts. Now measure the temperature of the water. It should have risen perceptibly, perhaps a half degree or more. The same temperature rise could have been obtained by bringing the bottle into contact with a hotter object, thereby, as is said, adding heat to it. But the end result, hotter water, would be indistinguishable from the water that was shaken. By measuring the temperature of the water, or any of its other properties (pressure and volume, for example), it is not possible to say how it got that way. This in turn implies that it is not possible to say which sample of water has more heat in it.

The *energy* of an object, however, is a well-defined property, much like its temperature, pressure, and volume. But energy is more abstract than heat, which, rightly or wrongly, we associate with temperature, something we can sense. Energy, like any abstract concept, acquires meaning from its attributes, not from its definition. The most important attribute of energy is that it is *conserved:* if an object is isolated from its surroundings, its energy is constant. This does not rule out the possibility that some of its other properties may be changing. Now suppose that an object is allowed to interact somehow with its surroundings. The two comprise the universe, which we assume does not interact with anything else because there isn't anything else. Thus the energy of the universe is constant, regardless of what its constituents are doing. If the energy of the object decreases, that of its surroundings must increase. In this instance we say that energy has been transferred from the object to its surroundings. And the reverse process is, of course, possible: the surroundings may lose energy to the object. Note that no palpable substance has been transferred in this process. Energy is not some kind of fluid that flows into and out of objects. Yet at one time heat was viewed as such a fluid, called *caloric*. The caloric theory of heat is officially dead but some of its concepts live on in such phrases as "the flow of heat." Indeed, it is often helpful to look upon energy (or heat) as some kind of fluid even though we know better.

Energy may be transferred to or from an object in two different ways: by heating it or by doing work on it. The former process occurs when the object interacts with its surroundings at a different temperature; examples of the latter process are compressing or expanding a gas and shaking a bottle filled with water. Heat transfer is therefore a particular type of energy transfer process, not a transfer of some substance or property called heat. Although both modes of energy transfer, heat and work, can occur simultaneously, we shall henceforth be concerned only with heat transfer.

The energy of objects made of many common substances is roughly proportional to their temperature, particularly over narrow temperature ranges around room temperature. But this is by no means necessary, and many examples to the contrary can be adduced: for example, as we shall see in the next chapter, the energy of a mass of ice is less than that of the same amount of water, yet when water freezes its temperature does not change. Even excluding such phase changes, energy is not always proportional to temperature. For our purposes, however, we shall assume that the energy of an object is directly proportional not only to its mass, which is true, but to its temperature as well, which is sometimes true.

Now we are able to interpret Newton's law of cooling. When an object cools, its temperature decreases with time. Because of our assumption above, this means that its energy also decreases with time at the same rate; thus energy is being lost from the object to its surroundings by the process of heat transfer. If the two are at the same temperature the heat transfer rate is zero; the greater their temperature difference, the greater the heat transfer rate. Indeed, the exponential decrease of temperature observed by Newton implies that at any instant the heat transfer rate is proportional to this temperature difference: in a given time interval, one second, say, an object cools; in the next interval of the same duration it cools less because its temperature during this interval is less than during the preceding interval; and so on for each succeeding time interval.

Newton's law of cooling is not a law in the sense that it implies inevitability like the law of conservation of energy, to which there are no known exceptions. Rather, it is an approximate relation with a limited range of applicability. Indeed, the notion of *the* temperature of an object makes sense only for one that can be characterized by a single temperature; this is certainly not true, in general, for very large objects: temperatures near their centers may be greatly different from those near their boundaries.

In the next section I shall show results of experiments which follow Newton's law of cooling quite well. But more important, I shall investigate how the proportionality between heat transfer rate and temperature difference depends on the characteristics of an object and the state of its surroundings.

COOLING CURVES

Cooling curves are easy enough to obtain: just fill a flask or a bottle with water heated to perhaps 50 or 60°C and then measure the temperature of the water at more or less constant time intervals; the thermometer must be fixed in place, by inserting it through a stopper in the flask, for example. If temperature is plotted on a logarithmic scale versus time on a linear scale then a straight line indicates adherence to Newton's law of cooling. Moreover, the slope of such a line gives the proportionality between heat transfer rate and temperature difference.

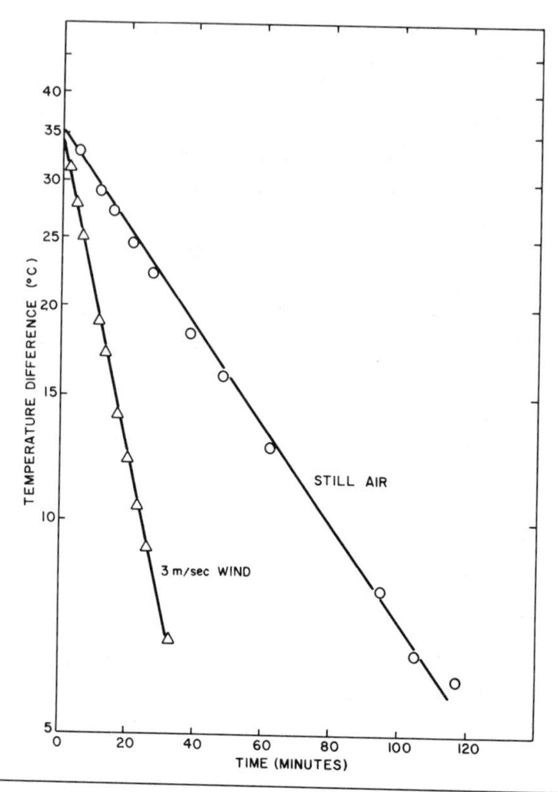

Figure 8.1 Cooling curves for a 125 ml flask of water in still air (circles) and in a 3 m/sec wind (triangles).

Cooling curves for a 125 ml flask filled with water are shown in Figure 8.1 for two different conditions of the surroundings: still air and air moving past the flask at a speed of about 3 meters per second (7 mph); the wind was generated by an electric fan. Note that nearly all the data points, circles and triangles, fall on straight lines. For ease of comparing experiments, I define the characteristic *cooling time* as the time required for the temperature difference to decrease by half. In wind the cooling time is 12 minutes; in still air it is almost four times greater, about 44 minutes. Engineers would say that in still air heat transfer occurs by *free,* or *natural, convection,* whereas in a moving air stream it occurs by *forced convection.* This distinction may seem quite artificial, if not meaningless, to meteorologists because winds of all scales in the atmosphere arise from temperature differences, not from the actions of pumps or blowers.

I called the air still but it really isn't, at least in an absolute sense. There are small, barely perceptible air currents around the flask. This is the mechanism for free convection: air in contact with the flask is heated, expands, and its density decreases, so it rises and is replaced by colder air from its surroundings. When air moves past the flask at a perceptible speed it whisks away energy at a greater rate, as evidenced by the shorter cooling time. This is consistent with our everyday experience of feeling colder in moving air than in still air at the same temperature.

Even in completely still air, heat transfer would occur by *conduction*. In fact, convection, which involves fluid flow, also involves conduction heat transfer through the thin film of stagnant fluid that adheres to objects; the nature of the flow determines the thickness of the film and, hence, the rate of heat transfer (for more on this see the next chapter). Conduction is the mode of heat transfer in solids; although there is no perceptible motion, there is motion at the molecular level. If a metal rod, for example, is colder at one end than at the other, its molecules at the hotter end are vibrating faster. They don't go anywhere but they do pass on their motion from the hot end to the cold end.

Still air is a poor heat conductor or, alternatively, a good insulator. But unless special care is taken, air moves easily, which greatly diminishes its value as an insulator. To exploit the insulating properties of air it must be confined in small spaces so that it doesn't move very much. This is the function of porous materials, such as cork and various foams, which are good insulators. Wool is

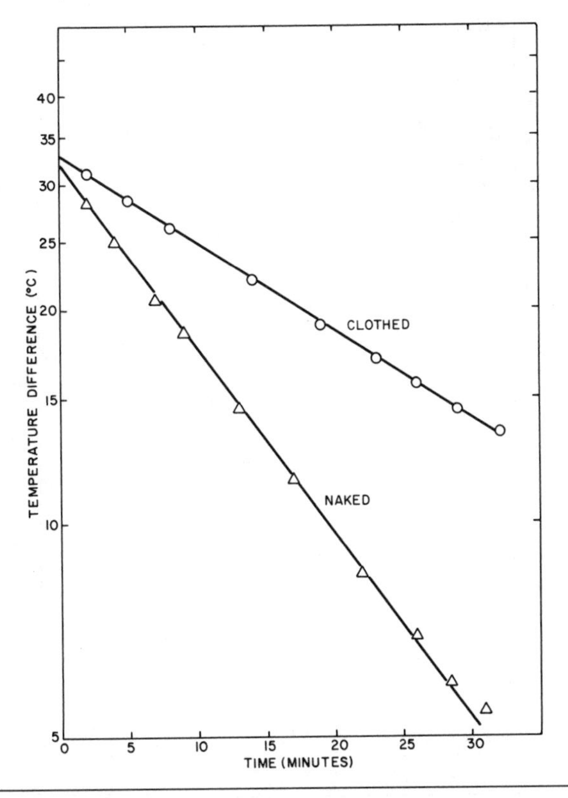

Figure 8.2 Cooling curves for a 125 ml flask of water in a 3 m/sec wind: naked (triangles); clothed with a thin layer of wool (circles).

another porous material. The effectiveness of wool as an insulator is shown in Figure 8.2. The cooling time for a naked flask in a 3 meter per second wind is 12 minutes; a modest covering of wool doubles this. Wool clothing is often said to be warm; but it could just as well be described as cool, for when wrapped around an ice cube it slows its melting rate.

Even if convection is suppressed, as in a gas at low pressure, an object can still exchange energy with its surroundings by *radiation* (I shall say much more about this in Chapter 10). The rate of radiation heat transfer depends on the temperature difference between the object and its surroundings as well as its composition and the condition of its surface. A metal such as aluminum, for example, generally radiates less energy per unit area than glass, both at the same temperature. This is particularly true if the aluminum is shiny. So wrapping aluminum foil around a flask of heated water should affect its cooling curve. That this is so is shown in Figure 8.3: aluminum foil increases the cooling time of a 125 ml flask in still air from 44 minutes to about 67 minutes. In a moderate wind (3 m/sec), however, the proportional increase is not so great: about 40

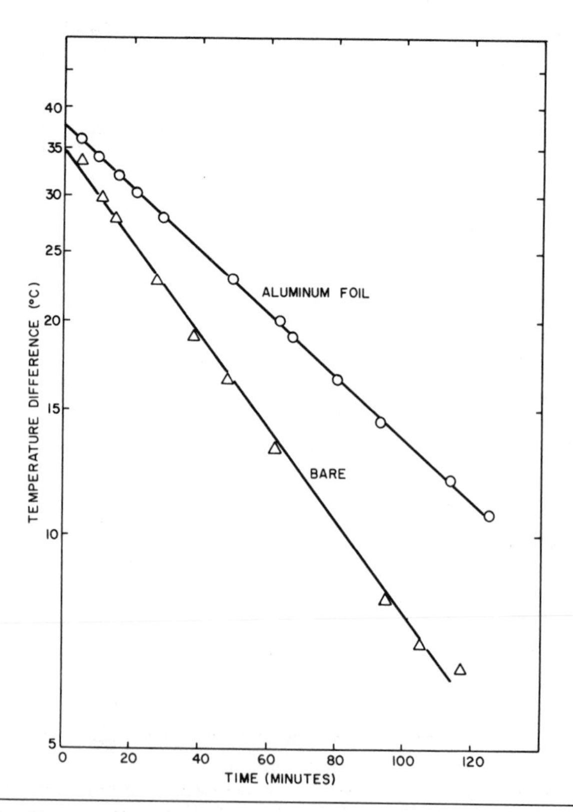

Figure 8.3 Cooling curves for a 125 ml flask of water in still air: bare (triangles); covered with aluminum foil (circles).

percent instead of 50 percent. So the relative importance of each mode of heat transfer (conduction, convection, radiation), all of which may be occurring simultaneously, depends very much on circumstances.

WATER AS A CONDUCTOR

Up to this point I have considered only cooling curves for flasks in air. When water is substituted for air the effect on cooling curves is dramatic. This is shown in Figure 8.4; the cooling time for a 1000 ml flask of water in still air is about 103 minutes, but in still water it is 8 minutes. This is in part because water is a much better conductor than air. The consequences of this great difference between cooling in air and in water are manifold. You can stand around naked all day—if that is what you like to do—in still air at 20°C (68°F) without discomfort. But try the same thing in water at the same temperature. Fully clothed, you wouldn't find an air temperature of 10°C (50°F) especially chilly. But if

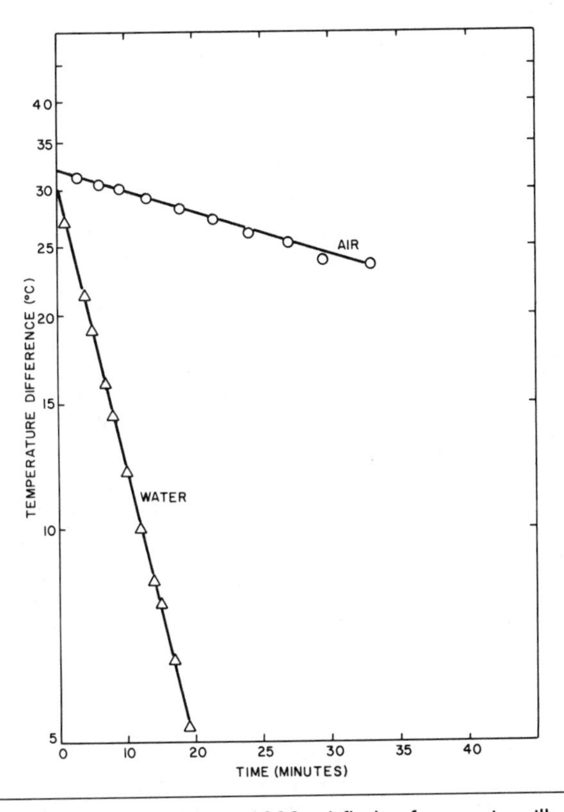

Figure 8.4 Cooling curves for a 1000 ml flask of water in still air (circles) and in still water (triangles).

you were to fall into water at this temperature, you could not expect to live long, perhaps an hour or two at most, unless rescued promptly. Unlike the flask I have used, the live human body is not a passive object: energy is generated within it by metabolic processes. But this may not be sufficient to compensate for the very high rates of energy transfer in water. A person whose internal body temperature falls much below 27°C (81°F) is not likely to survive the experience.

It is instructive to compare cooling curves for the 1000 ml flask (Fig. 8.4) and the 125 ml flask (Fig. 8.1), both in still air. The mass of water in the former is eight times that in the latter. You might guess, therefore, that the cooling times should stand in this ratio. But the cooling time for the 1000 ml flask is only somewhat more than twice that for the 125 ml flask. Why?

Energy is indeed proportional to mass or, if density is constant, to *volume.* But energy transfer occurs at the surface separating an object from its surroundings and is proportional to the *area* of this surface. Thus the rate of cooling is proportional to the *surface-to-volume ratio,* which means that the cooling time is inversely proportional to this ratio. Area has dimensions of length squared; volume has dimensions of length cubed; their ratio is therefore inversely proportional to a *characteristic length* (for a sphere, its radius), which we may take to be the cube root of the volume. The cube root of 1000/125 is 2, slightly less than the ratio of cooling times for the two flasks. To explain this small discrepancy would take us too far afield. Suffice it to say that, in this instance, the ratio of cooling times is determined primarily by the relative surface-to-volume ratios.

GUILTY OR NOT GUILTY?

Cooling curves did play an important role in the murder in Ceylon, the backdrop for this chapter. Here are Sir Sydney's words on what happened:

> The experiments on the cooling of recently dead bodies that I had suggested had been carried out. Professor de Saram [a former pupil of Sir Sydney's] had been fortunate in obtaining three bodies of young murderers immediately after they had been executed. Dressed in petticoats and sarees, the bodies were kept under conditions as close as possible to those in which the body of the murdered woman had been placed, and their temperatures were taken every half-hour. Approximately seven hours were required for these bodies to lose 5.2 degrees. This information was of crucial importance, for it was obvious that the much lighter woman would have taken less time to cool to the same extent; and, therefore, she would have died much later than 10.30 a.m.

This evidence seems to have tipped the balance in favor of the accused and after a three-month trial the jury pronounced him "Not Guilty." Who did it, then? It wasn't the butler—they didn't have one. It seems likely that it was the cook, but as far as I know, he was never brought to justice.

9

The Freezing of Lakes

The great tragedy of Science—the slaying of a beautiful hypothesis by an ugly fact.
T. H. Huxley

As I have moved from Arizona to Pennsylvania and then to New Hampshire, the word winter has become increasingly less abstract and depressingly more concrete. Regular occurrences which for Easterners are commonplace, are for me still a bit unexpected, even unnatural. For example, I used to think that freezing rain was a meteorological pipe dream, something that could occur in theory but rarely did in practice. It took only a few collisions with glazed sidewalks to disabuse me of this notion.

When winter arrives, various events not fully a part of my being begin to occur, such as the freezing of lakes. Those of my boyhood were properly fluid throughout the year, but those of my middle age acquire a coating of ice every winter. Yet even in the coldest of winters, the ice does not get very thick, certainly not as thick as might be expected from making ice cubes. Only a few hours are needed to produce an ice cube several inches thick in the freezing compartment of a refrigerator. On this basis, about a day would be needed to produce ice a foot thick; after a month, the ice would be about as thick as a three-story building. This is contrary to what is observed. Lakes do not freeze to astonishing depths when temperatures are as low as or lower than those in a freezer for several months. Even in the polar regions, ice does not become very thick. In his *Farthest North,* the famous Norwegian explorer Fridtjof Nansen gives a thickness of 11.5 feet for ice surrounding his ship "Fram" in the depths of the polar winter at about 80 degrees north latitude.

Extrapolation—the assumption that things will always change as they do now—is usually risky. My unrealistic prediction of the thickness of ice resulted from a *linear* extrapolation: I implicitly assumed that ice grows in equal thicknesses in equal time intervals. Yet it has long been known that ice should grow as the *square root* of time. The theory of this is hoary with age. It is usually associated with Josef Stefan, an Austrian, who in 1889 published a paper on a class of what have come to be called Stefan problems. Much earlier (1831), however, Gabriel Lamé, a French mathematician, and Benoit-Pierre-Emile Clapeyron, an engineer, had found that the thickness of ice is proportional to the square root of time.

I have known about mathematical Stefan problems for a long time. Indeed, I was so confident that ice should thicken as the square root of time, that I came close to not actually doing an experiment but merely suggesting that you do one. It always helps to check something for oneself, however, so I set myself the task of verifying what I knew to be true.

AN EXPERIMENTAL SURPRISE

Strictly speaking, the square-root law for ice freezing requires the zero of time to be taken when the liquid water is at its freezing point throughout. The way I did this was to put a container of water in the freezer until a thin ice film formed, then I broke the ice and stirred the ice-water mixture until most of the ice had melted. Then I began timing the growth of the ice.

After trying several schemes to measure its thickness, I finally adopted the rather crude technique of breaking the ice sheet, measuring its thickness, and then repeating the experiment for a longer time. In a finite container, ice does not grow uniformly but is thicker near the edges. To avoid this complication as much as possible I measured the thickness of the ice close to its center. Since I expected it to thicken as the square root of time, I elected to grow it for 1, 4, and 9 hours. The corresponding ice thicknesses would therefore be in the ratio of 1 to 2 to 3. After one hour, the ice was about 2 mm thick. In four hours, however, it was about 8 mm thick, nearly four times thicker instead of two. After nine hours it was 22 mm thick, about eight times thicker than it had been after one hour. Things were not going well. I did a bit of head scratching, a lot of cursing, and repeated the experiment with different containers. It must be the finite size of the container, I thought, although a few back-of-the-envelope calculations indicated that this was probably not the source of the problem. I tried better insulation. Shoving frozen meat and vegetables aside, I stuffed a container wrapped in thick foam into the freezing compartment of our refrigerator. I tried a polystyrene container. All with the same results. The ice was misbehaving. Fortunately, all this occurred during my sabbatical leave at Dartmouth, only a short distance from the U. S. Army's Cold Regions Research and Engineering Laboratory, where I had spent two summers. I called the resident expert on lake and river ice, George Ashton, and asked to see him, although I didn't say why. After I arrived at his office I didn't have to say more than a few words before he knew exactly what was troubling me. It was as if he had been waiting for me. He was quite familiar with my problem, had faced it himself, and had thought about it a lot. Before I pass on what I learned from him, I shall back up a bit and explain why I expected the ice thickness to increase as the square root of time.

THE FREEZING OF HYPOTHETICAL LAKES

The classical Stefan problem is as follows. A large body of water is at its freezing temperature throughout. Suddenly, the temperature at the upper surface drops below the freezing temperature. Ice begins to form. How does its thickness increase with time?

A given mass of ice is in a lower energy state than the same mass of liquid water, both at the same temperature. The salient characteristic of energy is that it is conserved: if any object is isolated from its surroundings its energy remains constant. When water freezes, therefore, it must interact with its surroundings since its energy does not remain constant. The energy of the water decreases while that of its surroundings increases by the same amount. We say that the water gives up its *latent heat* upon freezing.

We owe to Joseph Black, a professor of chemistry at Edinburgh, the first clear exposition of latent heat. His *Lectures on the Elements of Chemistry* was published in 1803, four years after his death. But the following statements had been made nearly forty years before: "When we deprive . . . a body of its fluidity . . . a very great quantity of heat comes out of it, while it is assuming a solid form, the loss of which heat is not to be perceived by the common manner of using a thermometer. . . . The extrication and emergence of the latent heat, if I may be allowed to use these terms," was to be distinguished from what he called *sensible* heat: latent heat "appears to be . . . concealed within the water." These archaic and even perhaps misleading terms, latent and sensible heat, are two of the many survivors from the discarded caloric theory of heat, in which heat was looked upon as a palpable substance capable of flowing from one place to another much like a fluid (see the previous chapter for a discussion of this and other related points). Indeed, everyday language and even that of modern theories of heat transfer, as opposed to their mathematical formulation, is consistent with the caloric theory. We commit no error by referring to the *flow* of heat, although we should recall, however dimly, that in so doing we are using a metaphor. Metaphors are useful, they evoke concrete images, but they are also dangerous in the minds of the careless (for more on metaphors in science, I recommend an article by Joanne Silberner in *Science News,* 18 October 1986).

We are agreed that when water freezes, energy must be transferred from it to its surroundings. Nothing has been said, however, about *how long* this might take. This depends on the nature of the surroundings. If they are solid (or at least sensibly motionless), the energy flow (or heat flow) occurs by conduction. The fundamental law of heat conduction is that annunciated by Jean-Baptiste-Joseph Fourier in his celebrated *Analytical Theory of Heat* published in 1822. I say "law," implying inevitability, but it would be more correct to call it a very good approximation. According to Fourier, the rate of heat flow at any point of a substance is proportional to the temperature *gradient* at that point; the constant of proportionality, called the *conductivity,* being different for different substances. The concept of a gradient is familiar to anyone who drives: 4% GRADIENT AHEAD. By this is meant that in every 100 feet, the road

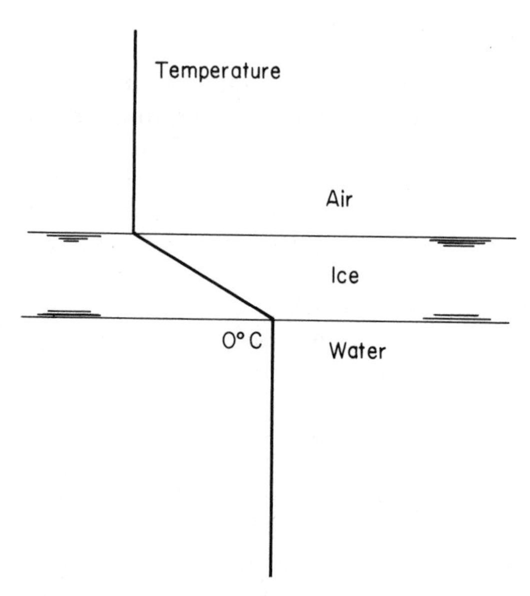

Figure 9.1 Ideal temperature profile. Initially, liquid water is at the freezing point throughout. Suddenly the air temperature drops below freezing. Under the assumption that the surface of the ice is always at the air temperature, the thickness of the ice increases as the square root of time. Thus the rate of growth of ice is smaller the thicker it becomes.

ascends (or descends) 4 feet. So also is it with temperature gradients: they are temperature changes over a given distance.

Now we have all the ingredients at hand to discuss Figure 9.1. At the ice-water interface, the temperature is 0°C (32°F); at the air-ice interface it is, by supposition, less than 0°C. As water below the ice freezes, it gives up its latent heat, which must be conducted through the ice above it. The rate at which this occurs depends on the temperature gradient in the ice. For a fixed temperature difference between the upper and lower surfaces of the ice, this gradient decreases with increasing ice thickness. As the ice thickens the gradient decreases, hence the rate of freezing of still more ice decreases. So we do not expect the thickness of the ice to increase linearly with time. This would occur only if the temperature gradient were constant.

There is another way to interpret the freezing of ice. The ice can be considered as a *resistance* to the flow of heat, in analogy with electrical resistance to the flow of electrical current. The thermal resistance of any slab of material is directly proportional to its thickness and inversely proportional to its conductivity. As the ice thickens, its resistance increases, and hence it better insulates the liquid water below it. Ultimately, this is why lakes do not freeze to astonishing depths.

I have argued against a linear increase in ice thickness with time. Yet this is what I measured. And when theory and experiment disagree, it is the theory that must be judged the loser. On the other hand, it is a matter of common experience that lakes do not freeze to the thicknesses that would be predicted on the basis of my experiments. To reconcile these apparent contradictions between theory, experiment, and observation, we need to take a closer look at the freezing of water.

THE FREEZING OF REAL LAKES

When trying to reconcile the differences between what one expects and what one observes, it is best to examine one's assumptions, those that are explicit, and even more importantly, those that are implicit.

Without giving it much thought, I implicitly assumed that the temperature at the upper surface of the ice is always that of the surrounding air. What really happens is shown in Figure 9.2. Just above the surface there is a thin layer of still air—the thermal boundary layer—over which the temperature decreases rapidly. The temperature gradient just outside the ice is much greater than that in it because the conductivity of air is about 100 times less than that of ice. Not only is the temperature at the upper surface *greater* than that of the surrounding air, it is also changing with time. Thus in the previous instance the temperature gradient in the ice is overestimated, hence so is the freezing rate.

In the initial stages of freezing, the surface temperature is decreasing. If all else were constant, the temperature gradient in the ice would increase. But all else is not constant: the ice grows, which results in a decrease of the temperature gradient. Recall that *two* ingredients go into a temperature gradient: a temperature difference and the distance over which it occurs. Initially, the temperature difference is decreasing at the same rate that the ice is thickening. As a consequence, the temperature gradient is constant and the ice grows linearly with time. But this does not prevail indefinitely.

To proceed, it may be easiest to use the language of thermal resistances, the "R-values" familiar to home owners and builders. The rate of heat flow from the water to the surroundings, driven by a temperature difference between the

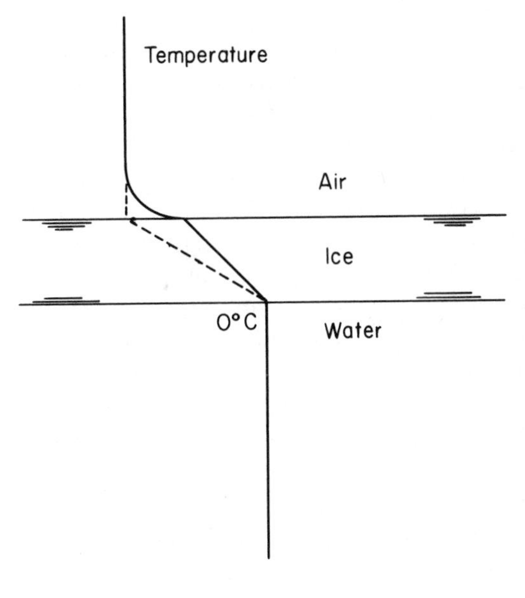

Figure 9.2 Real temperature profile. The surface temperature of ice overlying liquid water is always greater than that of the air. As a consequence, the temperature gradient in the ice is always less than it would be if the surface temperature were equal to the air temperature (dashed line). Initially, the ice thickness increases linearly with time because the surface temperature is decreasing at the same rate that the ice thickness is increasing. But as the surface temperature approaches a constant value, the ice grows as the square root of time. The temperature profile in the ice is taken to be linear, which is an approximation.

air and the water, is regulated by two thermal resistances in series: those of the ice and of the thermal boundary layer. The resistance of the boundary layer is constant (for a given wind speed), but that of the ice increases with time. At some point the resistance of the ice will dominate the total resistance (the sum of the two) and that of the boundary layer will be negligible. When this occurs, the thickness increases (approximately) as the square root of time. Initially the ice grows linearly with time (what I observed but did not expect), but eventually grows as the square root of time (what I expected but did not observe). Although there is no abrupt transition, the ice on lakes typically grows approximately as the square root of time when it is about 30 cm thick, according to George Ashton. Unfortunately, this large thickness means that it would be difficult to observe the square-root law of ice growth in a refrigerator. I say difficult but not impossible. One could direct a stream of air over the water. The thickness of the thermal boundary layer decreases with increasing wind speed. If we could mount a high-speed fan in the freezing compartment of a refrigerator, then we might be able to decrease the resistance of the boundary layer to a negligible value. It would take a powerful fan, however, one capable of producing wind speeds of over 100 mph. "Not in my refrigerator," my wife says.

Since thermal resistance is proportional to thickness and inversely proportional to conductivity, the resistance of an ice sheet 10 cm thick is the same as that of a still air layer 1 mm thick. If the resistance of an ice sheet 30 cm thick is much greater than that of the boundary layer, it follows that the thickness of the latter is of order 1 mm.

A SUGGESTED OBSERVATION

To observe the transition from linear to nonlinear growth of ice, one must abandon the comfort of a warm kitchen and venture out onto a frozen lake in winter. It is not just time that determines the thickness of ice on a lake, but the product of time and temperature below freezing. Although this is strictly true only if the temperature is constant, we nevertheless expect the ice thickness to increase with the number of *degree days* below freezing (the cumulative product of time, measured in days, and temperature below freezing). If you collect weather data and have access to lakes or ponds that freeze during the winter, you can do something with this data. I suggest that you observe the growth of the ice throughout the winter, and combine your observations of ice thickness with those of degree days below freezing. After a few seasons you should have a pretty good idea of how lakes freeze and, indeed, will have a means for predicting ice thicknesses.

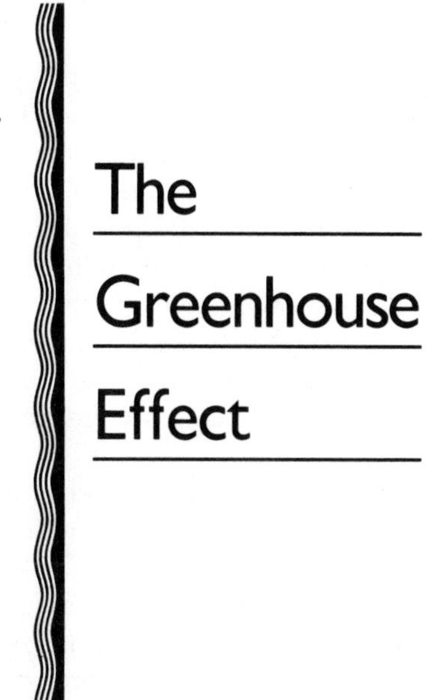

10 The Greenhouse Effect

Fear no more the heat o' the sun
Nor the furious winter rages
William Shakespeare: *Cymbeline*

Hardly a month goes by without headlines trumpeting imminent catastrophic global warming—what has come to be called the greenhouse effect—because of ever-increasing concentrations of atmospheric carbon dioxide. Some appeal to our conscience: U.S. URGED: HELP EARTH FROM GETTING TOO WARM. Others are merely ominous: HOW POLLUTION SWELLS WORLD OCEANS. All are intended to give us the sweats at night. As might be expected, hastily written books are pouring out in the wake of all this free publicity: cataclysm is one of America's fastest growing industries. A novel has been based on the theme of global warming; soon we can expect a Broadway musical, even a movie—with Raquel Welch playing the part of a carbon dioxide molecule. But as often happens, competition looms on the horizon. "Nuclear winter" has already begun to eclipse the greenhouse effect as the most-written-about, most-worried-about global catastrophe. This chapter may therefore be out of date by the time it is published. Nevertheless, I offer the following for those who would like to understand what underlies all the greenhouse-effect hoopla.

INVISIBLE RADIATION

We are almost blind. Although bathed in all kinds of electromagnetic radiation, we see only a small part of it, that with wavelengths between about 400 and 700 nm (1 nm = 1 nanometer, which is a billionth of a meter); the precise limits of this interval are of less importance than that it is exceedingly small compared with the enormous range of possible wavelengths. Yet expeditions beyond this interval were not undertaken until 1800. William Herschel led the way.

Although Herschel's life-long passion was astronomy, he made his living as a music teacher until his discovery of the planet Uranus propelled him into a fame that gave him the financial independence to pursue astronomy full time. As part of an investigation of the distribution of energy in sunlight he passed it through a prism, and with sensitive thermometers determined the relative heating powers of light of different wavelengths. The heating power increased in going from violet to red. Beyond the red, whence we get the name *infrared,* where no light was perceived, the heating power was greater yet. Thus Herschel found evidence for invisible radiation with a heating power greater than that of visible radiation.

I occasionally toy with the idea of trying to repeat Herschel's experiment. In the end, laziness dictates a less ambitious one, which I borrowed from Alistair Fraser.

Paint the bulb of an inexpensive thermometer black—an aquarium thermometer will do—and attach two binder clips to it as supports (see Fig. 10.1). When this thermometer is placed on the plate of an overhead projector the liquid column will slowly ascend because radiation from the projector is absorbed by the black paint. Note the length of the liquid column after it ceases to rise. Then place a shallow transparent dish filled with water under the bulb. Now the liquid column will slowly descend.

Except in thick layers, meters or more, pure water is transparent to *visible* radiation. But this does not imply that it is transparent to *invisible* radiation. Our eyes are as incapable of judging the transparency of water to invisible radiation as our ears are of judging the merits of an inaudible symphony. Observations of invisible radiation must necessarily be indirect.

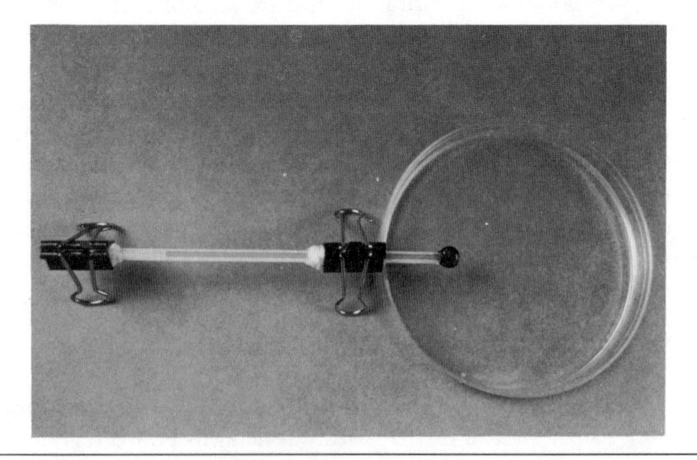

Figure 10.1 Evidence for invisible radiation. Light from an overhead projector is absorbed by a thermometer with its bulb blackened. When the light is transmitted to the bulb through a shallow layer of pure water, the thermometer registers a lower temperature.

A possible interpretation of the experiment is that the projector emits radiation we cannot see but which nevertheless can be absorbed by objects thereby heating them. And water is transparent to radiation we can see but opaque to that we cannot see.

Water is indeed an infrared filter. If you peer into the innards of a slide projector you'll find another such filter, a slab of glass next to the lens. You may have seen this before and wondered what purpose it serves: the projector works just as well without it. This glass slab is transparent to visible radiation but more or less opaque to infrared radiation. Only visible radiation is needed for projecting slides; infrared radiation is not only useless, it may burn them. Hence the unneeded component of the projector's radiation is filtered out to protect slides.

I have skirted perilously close to conveying the erroneous notion that there exists something called "heat radiation," special radiation with the unique property of heating objects that absorb it. No such radiation exists.

It was once believed that three distinct kinds of radiant entities existed: actinic, luminous, and thermal. The first was so named because of its ability to cause chemical reactions, the blackening of photographic plates, for example; these days we would call it ultraviolet radiation. The second was merely visible radiation. The third was what we would now call infrared radiation. As long ago as 1889 Lord Rayleigh asserted that he "certainly never believed in the three entities." Moreover, he quotes from Thomas Young's much earlier (1807) lectures that these radiations are "distinguished from each other . . . not by natural divisions but by their effect on our senses." Young states further that radiations "when sufficiently condensed, concur in producing the effects of heat." This is a clear statement that any sufficiently intense radiation is capable of heating.

Rayleigh goes on to say that "during the decade 1850–1860 all the leading workers in physics . . . held the modern view of radiation. It would be quite consistent with this that . . . workers in other branches of science, who trusted to more or less antiquated text-books for their information, should have clung to a belief in the three entities."

Rayleigh would find today's textbooks still antiquated. Although their authors rarely mention actinic rays, they cling to their belief in heat rays. Those who cherish this belief should have the courage of their convictions. Let them meditate on their sins while curled up in a microwave oven. Microwave radiation is not heat radiation so it cannot harm them. Better yet, let them expose their backsides to visible light from a high intensity laser. This is not heat radiation either, so it too cannot burn the true believer.

These two examples show that the heating power of radiation depends not only on its wavelength but on its intensity and the nature of the object it interacts with as well. Liquid water strongly absorbs microwave radiation (as evidenced by sales of microwave ovens); some plastics do not. Many solids and liquids strongly absorb infrared radiation; salt does not. Soot strongly absorbs visible radiation; ice does not. One can always find substances that strongly absorb radiation of a particular wavelength, and with an intense source they can be heated to any desired degree. That is, under proper circumstances *all* radiation is heat radiation.

Not only is there no special kind of radiation that deserves to be called heat radiation, there is no such thing as heat, an argument I made in Chapters 8 and 9. As far as I am concerned "heat" is acceptable as a verb: "to heat" is unobjectionable shorthand for "to increase the temperature of," just as "to cool" is shorthand for "to decrease the temperature of."

"Heat" as an adjective, as in "heat transfer," to indicate a particular way of changing a body's energy, may perhaps be used with caution. But to use "heat" as a noun, other than metaphorically, strikes me as a sure route to fuzzy thinking. Unfortunately, metaphors are rarely identified as such. Consequently, they often come to be taken literally and eventually cause mischief. Thus heat becomes a palpable substance in the minds of many of those who use this term as a noun. This leads to nonsense such as the notion that heat rises (hot air may rise, but heat never).

When the notion of heat as a substance is combined with the notion that heat is radiated, further nonsense ensues. For example, if heat rises and if heat is radiation (I am not making this up, I encounter these notions—and worse—with disturbing regularity), then heat radiation can be transmitted only upwards, which is contrary to experience.

Lest you think that I am quibbling over minor points of language, I note that in my experience many of the misconceptions people harbor have their origins in imprecise language. George Orwell stated that the English language "becomes ugly and inaccurate because our thoughts are foolish, but the slovenliness of our language makes it easier for us to have foolish thoughts." He was referring to political thoughts, but his remarks apply equally well to scientific ones. Precise language is needed in science, not to please pedants but to avoid absorbing nonsense that will take years, if ever, to purge from our minds.

WHY AND HOW OBJECTS EMIT RADIATION

Now that we are aware of the existence of invisible radiation we must try to understand why and how it is emitted by objects. Consider a single small charge, an electron, say. It will repel a second electron in the direction along a line joining them. The closer they are to each other, the greater this repulsive force. We may consider an electron—or, indeed, any collection of charges—to modify its surroundings, that is, to set up what is called an *electric field*. The force on any charge brought to a point in this field is the product of the charge's magnitude and the strength of the field at that point. Since force has a direction, so also does an electric field.

An electric field is imperceptible: we can neither see it nor touch it; it is a construct of the mind. To visualize it we may imagine lines to be drawn parallel to it. These field lines will show us the direction of the field at a glance. Moreover, the field strength is greatest where the field lines are closest together. A simple

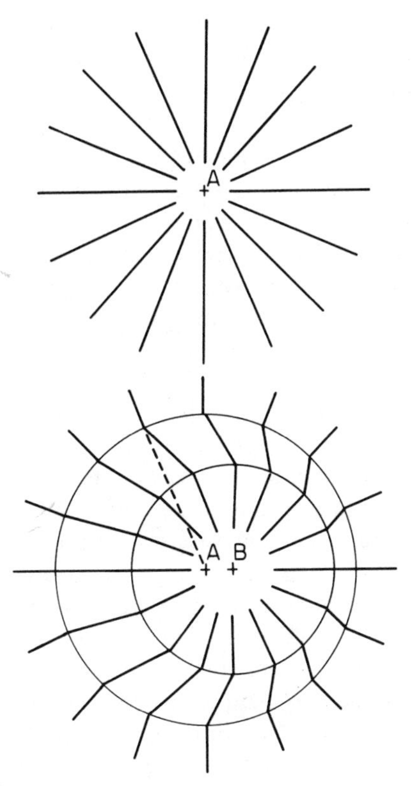

Figure 10.2 Around any charge we may imagine lines to be drawn parallel to its electric field. A few such field lines are shown for a charge at point A (top); for clarity, they have not been extended to A. When the charge is moved to point B (bottom), the field does not change instantaneously. Outside of a sphere the radius of which grows at the speed of light, the field is what it was before the charge was moved.

example of field lines, those surrounding a point charge, is shown in the top part of Figure 10.2. Note that the field lines are radial and closest together nearest the charge, which is consistent with my statements about the force exerted by one point charge on another. All the field lines are not shown, only enough to enable us to visualize the field.

Consider now what happens when a charge, initially at rest, is moved a short distance very rapidly and brought to rest again. The field lines change, but not instantaneously. It takes time for the message that the charge has moved to be propagated throughout space. Although the speed of propagation is very great it is finite—the speed of light. After the charge occupies its new position the field lines originating from it will again be radial, but only out to a distance equal to the speed of light multiplied by the time interval since the particle was again at rest. Thus within a sphere the radius of which grows at the speed of light, the field lines are those of the charge at its *new* position. But outside a sphere centered on the original position and with radius equal to the speed of light multiplied by the total time interval since the charge was moved, the field is that of the charge at its *original* position. Outside of this sphere the message that the charge has moved has not yet been received. Between these two spheres the field lines have kinks in them, which are sharpest for field lines perpendicular to the displacement and smoothest for those parallel (see Fig. 10.2). These propagating kinks are roughly analogous to those in a stream of water from a garden hose that result from moving it suddenly.

Because the charge was moved a short distance from rest and brought to rest again (i.e., it was *accelerated*), an electromagnetic pulse—a moving kink—was radiated outward at the speed of light. The strength of this pulse, its *amplitude*, is greatest perpendicular to the charge's displacement. Why *electromagnetic?* Magnetic fields are produced by currents, charges in motion; for example, a magnetic field surrounds an ordinary wire carrying an electric current. While the charge is moving it is a small current, and a magnetic pulse therefore accompanies the electric pulse.

Now imagine the charge (or the garden hose, to continue with the analogy) to be moved back and forth continuously; the number of return trips it makes each second is called its *frequency* of oscillation. Instead of a single pulse, a continuous string of them, an electromagnetic wave similar in many ways to water waves, is generated. The greater the frequency of this wave the shorter its *wavelength*, the distance between successive peaks. Thus by moving a charge back and forth at different frequencies we can generate the entire electromagnetic spectrum, from radio waves to gamma rays, with microwave, infrared, visible, ultraviolet, and X-radiation in between, an enormous range of frequencies. These designations are to some extent arbitrary: there are no sharp boundaries between the different kinds of electromagnetic waves. They differ by frequency (equivalently, wavelength), which varies continuously.

Although all electromagnetic waves are similar, they may be generated by different means. For example, radio waves are often generated by wires carrying electric currents (antennas), whereas gamma rays are generated by nuclear distintegrations. Yet even classifying waves by how they are generated is somewhat arbitrary: we may consider all of them to be generated by antennas of different size: radio waves by large antennas, microwaves by somewhat smaller ones (you have probably seen both kinds); infrared, visible, and ultraviolet waves by antennas with atomic dimensions; gamma rays by antennas with nuclear dimensions. Thus the size of an antenna may vary from many meters or more to billionths of a meter or less.

Electromagnetic waves also differ in the degree to which they interact with matter. If any source of radiation may be looked upon as a transmitting antenna (transmitter), the matter the radiation interacts with may be looked upon as a receiving antenna (receiver). The essential difference between receivers is merely the frequency of the waves they respond to. When you turn the dial of a radio you are adjusting its circuits so that they are in tune with the frequency of the radiation transmitted by the station whose broadcast you want to hear. And so it is with any electromagnetic radiation, not just radio waves. A receiver in tune with radiation from a transmitter will respond strongly to that radiation; the receiver may be of molecular, atomic, or even nuclear dimensions.

A telephone sits on my desk. To my crude senses it appears to be stationary.

And yet it is alive with activity. It is like a swarm of bees, which appears motionless until one takes a closer look.

Each molecule of my telephone moves now in one direction, now in another, going nowhere on average, just as I paced back and forth while writing this, going nowhere literally and sometimes figuratively. Although matter is usually electrically neutral, it is nevertheless composed of electrical charges—negative electrons and positive nuclei in equal amounts. These charges are the building blocks of atoms, which may group into more or less stable molecules. Thus my telephone—indeed, all matter—radiates electromagnetic energy because it is composed of oscillating charges. Such radiation is not confined to a single frequency or even a narrow range of frequencies. Because of the different motions of the vast number of molecules composing any object, it emits radiation of all frequencies.

This is not to say, however, that radiation of all frequencies is emitted uniformly. The distribution of radiation emitted by an object—its *emission spectrum*—depends on its composition. More important for our purposes here, this spectrum depends strongly on temperature: the greater the temperature, the greater the average frequency of the emitted radiation. Before discussing a demonstration of this, I want to dispel a widespread misconception.

All objects at temperatures above absolute zero (−273.16°C) emit radiation of all frequencies. But this does not imply, as is sometimes stated, that all motion stops at absolute zero. I believe that this misconception stems from the behavior of *ideal* gases.

An ideal gas is one composed of molecules so far apart that they do not interact with one another; at normal temperatures and pressures, air is approximately an ideal gas. The absolute temperature of such a gas is proportional to the average kinetic energy (i.e., energy of motion) of its molecules. So it is indeed true that motion of *ideal* gases ceases at absolute zero, which is largely irrelevant to what happens to *real* gases: at very low temperatures ideal gases do not exist. All gases become liquids at sufficiently low temperatures; at lower temperatures still, they solidify. And a solid is definitely not an ideal gas; its constituent atoms are so closely packed that they interact strongly with one another.

The electrons of an atom orbit its nucleus like planets around the sun. If it were true that *all* motion came to a halt at absolute zero, we would have to conclude that all matter collapses at this temperature; the negative electrons would cease to orbit the positive nucleus, they would be attracted toward it, and the atom would collapse. Even if we ignore the electronic motion, it is still not true that the atoms of a solid become motionless at absolute zero. Although their energy is a minimum, they still partake of what is called the *zero-point motion*. To understand this, however, requires us to abandon the familiar ideas of classical physics. The behavior of matter at very low temperatures is explicable only by quantum physics, which is well beyond the scope of this book. I merely point out that although matter ceases to radiate at absolute zero, its constituents do not cease to move.

AN EXPERIMENT TO BE DONE AT LEISURE

Perhaps without realizing it, you have undoubtedly observed the shifting of an object's emission spectrum toward higher frequencies (shorter wavelengths) as its temperature is increased. Consider, for example, what happens when you turn on an electric stove in darkness. At first, its elements do not emit enough visible radiation to be seen. As the temperature of an element increases, however, it will gradually become visible. Not only does it emit more radiation of all frequencies, it emits a greater fraction of visible radiation. A photograph of the element of a hot plate is shown in Figure 10.3. I took a five-second exposure

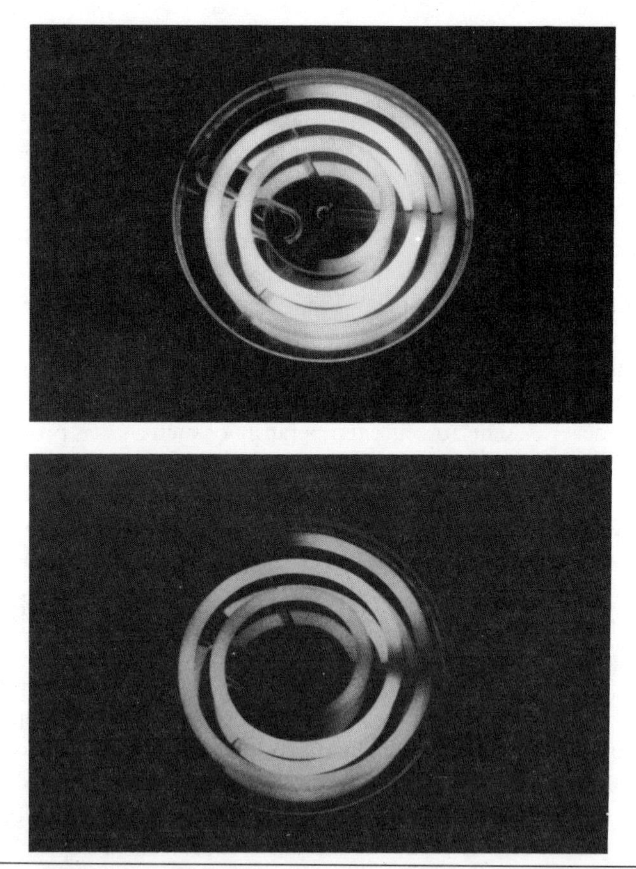

Figure 10.3 The top photograph was taken in darkness with the heating control at its highest setting; the aperture was f/3.5 and the exposure time was 5 seconds. The bottom photograph is of the same heating element but with the control turned down to the point where the element was no longer visible; the exposure time was 12 hours. ASA 1000 film was used for both photographs.

using ASA 1000 color film, the new sensitive film for low light levels, in the darkest corner of my basement with all lights turned off and the windows covered. To provide evidence that objects you cannot see really do emit visible radiation I turned down the heating control until the element was no longer visible, although it was still too hot to touch. Then I took another photograph (also shown in Fig. 10.3), but this time I exposed the film for 12 *hours* instead of for 5 *seconds*.

We can see in an instant feeble light that will not blacken even the most sensitive film—unless it is exposed for a long time. Although we do not see objects as brighter the longer we look at them, the brightness that film records depends on exposure time. When the element was hottest it was emitting perhaps 10,000 times as much visible radiation as it was when it was colder. I could see this element in darkness more or less instantaneously, yet to photograph it I had to expose the film for 5 seconds. When I turned the heating control down, the element became invisible to me no matter how long I stared at it. But by increasing the exposure time by a factor of 10,000, I was able to photograph what I could never see.

The colors of the two photographs were different. Short exposure of a hot element gave a yellowish-orange color, whereas long exposure of a colder element gave a reddish color. This is as expected: the higher the temperature the higher the average frequency of the emitted light, and both orange and yellow light correspond to frequencies higher than those of red light.

BLACKBODY RADIATION

Although the hotter an object is the more radiation it emits and the higher the average frequency of this radiation, emission spectra of different objects at the same temperature may be quite different. There is, however, a class of hypothetical objects, called *blackbodies,* with the property that their emission spectra depend only on their temperature. As its name implies, a blackbody absorbs all incident radiation regardless of frequency, direction of incidence, and even polarization state (see Chapter 19 for a discussion of polarization). The term blackbody, I have discovered, leads to confusion, especially when accompanied by vague statements about how a good absorber is a good emitter. What is "good" depends on one's point of view; to a designer of solar collectors, a good emitter is one that does not emit at all. Moreover, when told that a black object—my telephone, for example—is a good emitter, people are puzzled by why, to their eyes, it isn't emitting. Finally, I question the wisdom of introducing blackbody radiation by way of something nonexistent.

A term for blackbody radiation that gives a better clue to its origin is *equilibrium radiation*. Imagine a cavity carved out of any material, its walls opaque to all radiation, which can be obtained if they are thick. For definiteness, we may take them to be aluminum, which is certainly not a blackbody: aluminum, especially if polished, is a good mirror, not only for visible radiation but for radiation from ultraviolet to radio frequencies as well.

Radiation of all frequencies is emitted by the walls, as well as absorbed or reflected by them; these processes go on continually. When equilibrium has been reached (i.e., the temperature of the cavity's walls does not change with time), the spectrum of radiation within the *real* cavity is the same as that emitted by a *hypothetical* blackbody. To understand why, imagine a blackbody to be placed in the cavity; it is bathed in the radiation emitted and reflected by the walls. When the body's temperature ceases to change, the rate at which it emits radiation must be equal to the rate at which it absorbs the radiation incident on it. By definition, it absorbs all incident radiation. Constancy of temperature is ensured if the blackbody's emission spectrum is the same as that of the radiation it absorbs, which is the equilibrium radiation filling the cavity.

Although a flat piece of aluminum is not a blackbody, an aluminum enclosure fills with blackbody radiation because of multiple reflection of radiation emitted by its walls. Strict blackbodies do not exist, but blackbody radiation does exist—it is radiation in equilibrium with matter.

I must emphasize that all bodies, black or not, at temperatures above absolute zero, emit radiation of all frequencies, and I do so not because of extreme fastidiousness. Not long ago I received a NASA report by Alfred Chang in which he reviews *microwave* emission by snow. Although snow doesn't emit enough microwave radiation to cook steaks (this is fortunate, because if it did, we would also be cooked), it does emit measurable amounts. And this emission is being exploited as a means for remotely sensing snowpacks through the use of satellites.

All bodies do not emit uniformly; the frequency of peak emission depends on temperature. Emission spectra of bodies at terrestrial temperatures peak in the infrared. Consider, for example, the blackbody emission spectra shown in Figure 10.4 for temperatures of 250°K (−23°C) and 320°K (47°C). Blackbodies at temperatures within this range emit radiation of all frequencies, but their spectra peak in the 8–12 μm wavelength region (1 μm = 1 micrometer is a millionth of a meter). The total radiant energy emitted by any object is proportional to the area under its emission spectrum. For a blackbody, this is proportional to the fourth power of its absolute temperature. Thus a 320°K blackbody emits nearly three times as much radiant energy as a 250°K blackbody of the same size; both of them emit mostly infrared radiation.

I have switched abruptly from frequency to wavelength for convenience. It is equivalent to specify radiation by wavelength, although frequency is the more fundamental quantity.

Most terrestrial objects—sand, sea, forest, soils, even snow—are nearly blackbodies. Snow is the blackest of the lot. And it is also the whitest. Before you put this book down, muttering that dementia praecox finally has me fully in its grip, read on for a resolution of this conundrum.

We are, as I mentioned earlier, nearly blind, as a consequence of which we are optically parochial. To us, snow is white—when illuminated by white light,

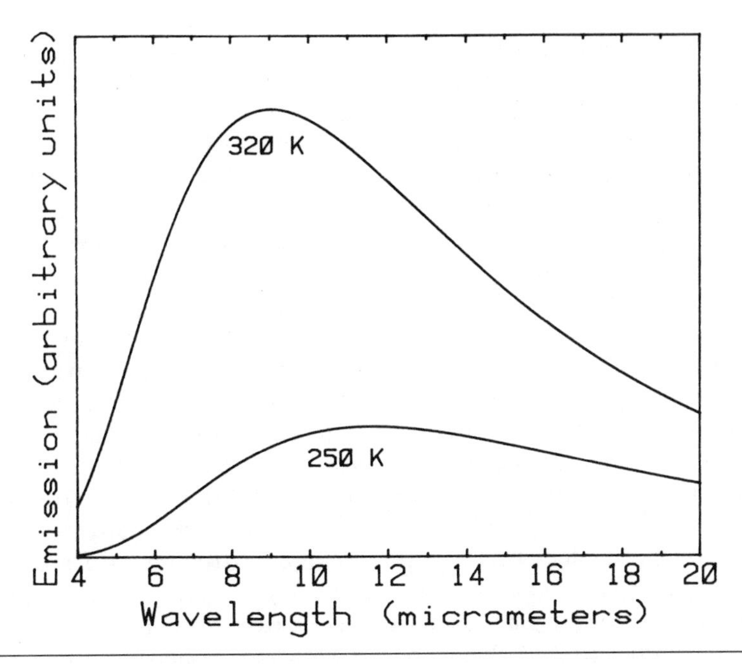

Figure 10.4 Blackbody emission spectra. Most terrestrial objects are nearly blackbodies. At normal temperatures their infrared emission spectra peak in the 8–12 μm region.

sunlight for example. Clean, fine-grained snow may reflect more than 94 percent of the *visible* radiation incident on it. Our eyes cannot tell us how much *invisible* radiation it reflects.

We would be extrapolating wildly by asserting that because snow is white (i.e., highly reflecting) at visible wavelengths, it is white at all wavelengths. Snow is black, as are most natural objects, to the kind of infrared radiation emitted by objects at normal temperatures. My statement that snow is the blackest and the whitest substance on our planet is indeed true, but it needs to be qualified to make sense, as do all statements about what is black and what is white.

My telephone emits radiation I cannot see. Because it doesn't reflect much radiation I can see, I call it black. By the same criterion–low reflection–it is also black at infrared wavelengths, but so are telephones I would call white. I have seen photographs, taken using infrared radiation, of people of different races. Between about 3 and 15 μm human skin is approximately a blackbody. All of us are black in the infrared, no matter how much we may glitter and gleam in sunlight.

I have been using the term infrared somewhat loosely. The infrared radiation I have had in mind is that emitted by objects at normal terrestrial temperatures. At the beginning of this chapter I sketched William Herschel's discovery of infrared radiation. This was not terrestrial infrared radiation, however. Herschel only skirted the vast kingdom of infrared, which extends from about 0.7 to 1000 μm. He discovered what we now call the *near* infrared (note the parochialism again: "near" means near to visible). Terrestrial radiation is called *middle* infrared by some folks, *far* infrared by others. By near infrared radiation I mean that shortward of 1.5 μm; by far infrared, that longward of 5.6 μm.

The sun emits infrared radiation, about half of its radiant energy. Most of the radiation emitted by the sun lies between 0.25 and 2.5 μm whereas most of that emitted by terrestrial objects lies between 4 and 24 μm. Because these two wavelength regions do not overlap, a distinction can be made between solar radiation (or shortwave radiation) and terrestrial radiation (or longwave or thermal radiation). Infrared radiation, without qualification, often denotes terrestrial radiation, but to avoid confusion it is best to state what wavelengths one has in mind. For example, infrared film commonly available is sensitive to near infrared radiation. But infrared satellite photographs may have been taken with radiation from near to far infrared. Near infrared photography exploits *scattered* solar radiation, far infrared photography exploits *emitted* radiation. The distinction is not trivial: the source of near infrared radiation largely vanishes with the setting sun whereas far infrared radiation is emitted day and night.

Because most terrestrial objects are nearly black to the kind of radiation they emit, their temperatures can be measured without touching them. If we could measure the ratio of radiation emitted by a blackbody at two (or more) wavelengths, say 6 and 8 μm (see the emission spectra in Fig. 10.4), we could infer its temperature. Devices to do this, called infrared thermometers, exist, although they are not cheap.

The preceding paragraphs may seem a rather long-winded preamble to the greenhouse effect, which I have not yet explained despite my promise to do so at the beginning of this chapter. I ask you to be patient. The greenhouse effect deserves more than the kind of superficial treatment commonly found in newspaper articles. There are so many misconceptions surrounding the physics of the greenhouse effect that I am taking great pains to dispel them. A sound understanding is not built on shaky foundations.

I have laid most of the foundations for the greenhouse effect. Now I shall finish them, add walls, a roof, and even a bit of decoration.

SOME OBSERVABLE CONSEQUENCES OF TERRESTRIAL RADIATION

Terrestrial radiation, although invisible, makes its presence known in many ways. For example, while walking to work on fall mornings, I sometimes see frost-covered lawns like that shown in Figure 10.5. Near the tree the grass is free of frost. The tree is not a kind of frost umbrella. It no more shelters grass from frost than it shelters it from air. Frost does not fall, it is ice that has condensed from atmospheric water vapor onto subfreezing surfaces. The absence of frost near the tree indicates that the grass is warmer there.

Figure 10.5 A frost-covered lawn on a fall morning. Around the tree the grass is frost-free because it receives more infrared radiation from the tree than grass in the open receives from the sky.

Grass emits infrared radiation, which causes it to cool. But it also absorbs infrared radiation from its environment, which causes it to warm. Whether there is net radiative cooling or warming depends on the balance between emission and absorption. Grass in the open receives less infrared radiation from the atmosphere than it emits to it; a clear sky radiates approximately as a 250°K blackbody. The tree, however, emits more than the sky, enough to keep the grass near the tree warmer than that away from it. Examples like this abound—if you know what to look for.

It is sometimes stated that terrestrial objects absorb radiation during the day and then emit it at night. This borders on fantasy. It implies that all objects are equipped with detectors to tell them when the sun goes down so that they can begin emitting. Infrared radiation is emitted ceaselessly, day and night, throughout eternity. And because daytime temperatures are usually *greater* than nighttime temperatures, emission rates are usually *greater* during the day than at night.

A tree in sunlight absorbs solar radiation. It also absorbs terrestrial radiation from its environment. At the same time, it emits terrestrial radiation the amount of which depends on its temperature. These processes are essentially independent of one another. Emitted radiation does not deflect incident radiation thereby preventing it from being absorbed. At night the tree no longer receives solar radiation, although it continues to both emit and receive terrestrial radiation. It would be incorrect, however, to state that this emitted radiation is that trapped during the day. Radiation is not like lobsters. Absorbed radiation vanishes, but not without a trace: its energy lives on in whatever absorbed it, and part of this energy may eventually be emitted as new radiation. Thus all objects are graveyards and nurseries for radiation, but never its permanent abode.

Because of the tree, grass beneath it is warmer than it would otherwise be, both day and night. We are not aware of this until warming by the tree is sufficient to prevent frost from forming where it would otherwise. This is best seen in the fall as nights begin to get colder. In warmer months, other examples of warming by infrared radiation may be seen, such as that shown in Figure 10.6. This is a photograph of a beach taken by Alistair Fraser. During the night dew formed on the sand hence darkened it (see Chapter 15 for why wet sand is darker than dry sand). But the sand is not uniformly dark. The crests of the sand ripples are darker than the troughs, as evidenced by the brightness differences. It is warm enough in the troughs that dew does not form there. Sand in the troughs

Figure 10.6 Early morning dew on a beach at the edge of Kootenay Lake in British Columbia. (Top) The sand is drier near the boat because it is warmed by infrared radiation emitted by the boat. (Bottom) On a smaller scale, the sand ripples show the same effects of infrared radiation. Because the sides of the troughs radiate to one another, they receive more radiation than the crests, which receive only radiation from the sky. Photographs by Alistair Fraser.

receives not only infrared radiation from the sky but also from the surrounding sand whereas sand at the crests receives only infrared radiation from the sky. Note also that the sand near the boat is brighter, hence drier, than sand farther away. Again, the sand receives more infrared radiation than it would otherwise because of its nearness to the radiating boat. These examples, the frost-free lawn and the patterns of dark and light on a dewy beach, are small-scale consequences of absorption and emission of terrestrial infrared radiation. I should now like to move on to an example at a much larger scale, that of the entire planet.

RADIATIVE EQUILIBRIUM OF THE EARTH

Our planet is continuously bathed in solar radiation. Although we who are confined to a fixed spot experience day and night, the earth does not. It is always day in the sense that the sun is shining on one half of the globe. Much of the incoming solar radiation, about 30 percent, is scattered back to space by clouds, atmospheric molecules and particles, and everything on the earth. The remaining 70 percent is absorbed, mostly at the earth's surface. This absorbed radiation gives up its energy to whatever absorbed it, thereby causing its temperature to increase. Because solar radiation is absorbed ceaselessly by the earth, its temperature should get higher and higher. It does not, of course, because, the earth also emits radiation the spectral distribution of which is quite different from that of the incoming solar radiation. The higher the earth's temperature, the more infrared radiation it emits. At a sufficiently high temperature, the total rate of emission of infrared radiation equals the rate of absorption of solar radiation. Radiative equilibrium has been achieved, although it is a dynamic equilibrium: absorption and emission go on continuously at equal rates. The temperature at which this occurs is called the radiative equilibrium temperature of the earth. This is not the temperature at any one spot or at any one time but merely the temperature a blackbody must have in order for it to emit as much radiant energy as the earth absorbs solar energy.

If we assume that the atmosphere is transparent to terrestrial infrared radiation, the radiative equilibrium temperature of the earth is about 255°K (−18°C). But if the average surface temperature of the earth were this low, I would doubtlessly not be here to write about it, nor would you be here to read about it. Something is missing. Our assumption about the transparency of the atmosphere to infrared radiation is suspect.

You may find it difficult to accept that something so tenuous as the atmosphere absorbs and emits infrared radiation. Yet I have repeatedly said that *everything* does, and I am not going to back down now. The atmosphere is indeed tenuous, but it is also very thick. If it were compressed into a layer with the density of water, say, its thickness would be almost ten meters. A layer of liquid water a few millimeters thick is quite opaque to terrestrial infrared radiation, so perhaps it is now not so implausible that the atmosphere can be nearly opaque to such radiation.

There is a persistent misconception that sunlight directly heats the air around us. Although it is true that there is some direct heating of air because it absorbs solar radiation, indirect heating is of greater importance: the ground absorbs solar radiation, is heated, and in turn heats the air in contact with it. During the day, air temperatures usually decrease with height; the air is warmest where it is closest to the warm ground. The reverse would be true if air were directly heated by solar radiation. At night the temperature profile tends to reverse. Air temperatures are lowest near the ground because the ground cools, hence so does the air closest to it.

I have emphasized that the ground emits radiation day and night, ceaselessly. Some of this infrared radiation is absorbed by the atmosphere because of which it is warmer than it would otherwise be. But the warmer the atmosphere the more it emits to the ground, hence keeping the ground warmer than it would otherwise be. This radiative exchange between the atmosphere and ground is similar to that between the sand and the boat shown in Figure 10.6, the consequences of which are readily apparent. The extent to which the atmosphere absorbs infrared radiation is of great importance in determining the temperature of the ground. If the atmosphere absorbed no infrared radiation, the ground would be intolerably cold, and so would the air in contact with it.

Not all atmospheric gases absorb infrared radiation to the same degree. Indeed, the most abundant ones by far, oxygen and nitrogen, are the least absorbing. Of far greater importance are water vapor and much less abundant gases such as carbon dioxide, ozone, and methane. Carbon dioxide, in particular, is the one frequently in the headlines. There seems to be little dispute that carbon dioxide concentrations in the atmosphere have been increasing because of increased burning of carbonaceous fuels such as coal and oil. At present, for every one million molecules in the atmosphere, about 340 of them are carbon dioxide (this is written 340 ppm, parts per million). To those who snort that 340 ppm of anything must surely be of no consequence, I recommend 340 ppm of arsenic in their coffee.

No one knows for sure what will be the consequences of ever-increasing concentrations of carbon dioxide. But some educated guesses have been made, the most popular of which is that doubling the concentration of carbon dioxide will cause surface temperatures to rise by 2 or 3 degrees C. This increase of temperature as a consequence of increased concentration of carbon dioxide has come to be called the greenhouse effect.

Although carbon dioxide gets the blame for the greenhouse effect, it has been pointed out by V. Ramanathan at the National Center for Atmospheric Research (NCAR) that carbon dioxide is only the tail that wags the dog. According to his calculations, the role of carbon dioxide is merely to increase ocean temperatures so that more water evaporates. It is the increased water vapor in the atmosphere that is responsible for most of the temperature rise.

There are many controversies surrounding the greenhouse effect, even one concerning its name, which I turn to next.

HOW DO GREENHOUSES WORK?

The atmospheric science community seems to be divided into two groups. In one group are those who, when hearing the words "greenhouse effect," roll their eyes, shake, foam, and turn a delicate shade of purple while sputtering, "Greenhouses don't work that way." In the other group are those who are quite content with the term greenhouse effect. Indeed, they would probably be just as content if it were called the outhouse effect. I can afford to have a bit of sport over this because at one time or another I have belonged to both groups.

As you might expect, the term greenhouse effect, used to describe increased temperatures because of increased atmospheric carbon dioxide, arises from some real or imagined connection with greenhouses. To some folks the interiors of greenhouses are warm because of (metaphorical) radiation trapping: the glass is transparent to solar radiation but opaque to infrared radiation. The glass absorbs infrared radiation from the underlying soil, is warmer than it would otherwise be, hence keeps the soil warmer than it would otherwise be. If this explanation is correct, it is reasonable to describe the effect of increasing carbon dioxide as a greenhouse effect.

Other folks, however, assert that radiation trapping has little to do with the interior warmth of greenhouses. They are merely shelters from the wind. That is, they suppress convective heat transfer (for a discussion of the various modes of heat transfer see Chapter 8) rather than radiative heat transfer. If this is correct, the atmospheric greenhouse effect is a misnomer.

Both explanations have their adherents. Both are reasonable. What is worse, calculations can support either of these two views. After years of meditation, I have come to the conclusion that when reasonable people hold diametrically opposite views on a scientific subject, it is often that both are right—and wrong. One's conclusions depend very much on one's assumptions, both stated and unstated.

Over ten years ago (1974) a controversy about the mechanism for greenhouse warming erupted in the pages of the *Journal of Applied Meteorology*. A parallel, but nearly independent, controversy was initiated a year later by Ronald Schwiesow (now at NCAR) in the pages of *Optical Spectra*. This led to a spate of letters in this journal and in *Science, New Scientist,* and *Popular Science*. I am grateful to Schwiesow for passing on to me his collection of greenhouse controversy papers to add to mine.

John Kessler, a physics professor at the University of Arizona, wrote an incisive analysis of the various claims and submitted it for publication. Kessler's manuscript was rejected (for reasons I can't remember), but I have kept it all these years, and upon re-reading it, I find that it makes as much good sense now as it did then.

According to Kessler, both sides to the controversy have merit, which ought to please everyone. The relative contributions of convection suppression and radiation trapping depend very much on the greenhouse and its environment. The thicker the glass and the stiller the air, the more important is radiation trapping. The greater the wind speed and the thinner the glass, however, the less important it is. This should perhaps come as no surprise. On a cold day in a howling wind you stay inside your home, not because it traps radiation but because it shelters you from the wind. Indeed, it is almost a truism of heat transfer that the relative contributions of the various modes depend very much on circumstances. In Chapter 8 I suggest some ways to demonstrate this.

One of the most interesting examples of the role of convection in greenhouse temperatures is in a paper by Kirby Hanson in the *Journal of Applied Meteorology* (1963, Vol.2, p. 793). He states that "during January 1961 the minimum temperature averaged 2.4 F *lower inside* than *outside*" a small polyethylene-covered greenhouse. Here is an example of the greenhouse effect in reverse: the greenhouse suppresses mixing of warmer surrounding air with that inside it.

The role of the thickness of the glass is perhaps not so obvious. The thicker the glass, the greater the temperature difference between its outer and inner surfaces. Heat transfer from the glass to its surroundings depends on the temperature difference between the outer surface and the surrounding air. The colder this surface is, the lower the rate of heat transfer, hence the hotter it is inside the greenhouse. If you find yourself under fire for using the term greenhouse effect as shorthand for what happens in the atmosphere you need merely retort that *your* thick-walled greenhouse is set in a very calm spot. On the other hand, if you either snicker or fume when you hear or see the term greenhouse effect, you can draw sustenance from the fact that greenhouses are often set in windy environments and have thin walls.

THE OTHER GREENHOUSE CONTROVERSY

There is only one sure way of determining whether increased atmospheric carbon dioxide will cause average global temperatures to rise. Construct two planet earths, identical in all respects except that on one carbon dioxide remains constant and on the other it increases with time. Then observe the temperature changes on both planets. This project, desirable though it might be, is something that even a NASA project manager would shrink from undertaking. So we have to rely on an inferior substitute: computer modeling.

When movable type was invented the printed word became sacred. It was difficult to accept that words so beautifully displayed could be devoid of sense. Now that we have lived with books for centuries they have lost some of their sacredness, and the printed word is viewed with more skepticism. But computers are a recent innovation, and the output that spews from them tends to be regarded with the same awe once given to the printed word. Yet no matter how large the computer, no matter how many thousands of acres of virgin forest are devastated to provide paper for its maw, the results it produces are no better than the ideas and data that went into their production.

The current conventional wisdom is that doubling carbon dioxide will cause global temperatures to rise 2 or 3 degrees C. A vocal minority, however, asserts that not only will the temperature not rise, it will decrease. All such predictions necessarily require that simplifying assumptions be made. The boundary between simplification and oversimplification is obscure, but woe to him who crosses it. That carbon dioxide concentrations in the atmosphere have been increasing seems incontrovertible. Only the consequences of this are in dispute. Why this is so is because of feedbacks that are either unknown or are swept under the rug. The best example of feedback that I have ever seen was a black box on the side of which there was a toggle switch and a sign: DO NOT TURN ON. Perhaps you could resist the temptation, but I couldn't. The result was, after some grinding and whirring from within the box, that the lid opened, a hand emerged, turned off the switch, and then withdrew back into the box. Turning on the switch resulted in a feedback that eventually turned it off. And so it is also with the atmosphere. Increasing temperatures as a result of increasing concentrations of carbon dioxide may cause something unforeseen to happen that will result in lower temperatures. One possibility, for example, might be that warming the oceans will result in more water vapor in the atmosphere and, as a consequence, more clouds. But this might lead to more solar radiation scattered to space, hence lower temperatures.

Because of feedbacks, some known, some unknown, some accounted for correctly, others incorrectly, the carbon dioxide controversy will continue to rage until the evidence is before our very eyes. I say our eyes, but I mean someone else's eyes. It is likely that neither you nor I will be around to see.

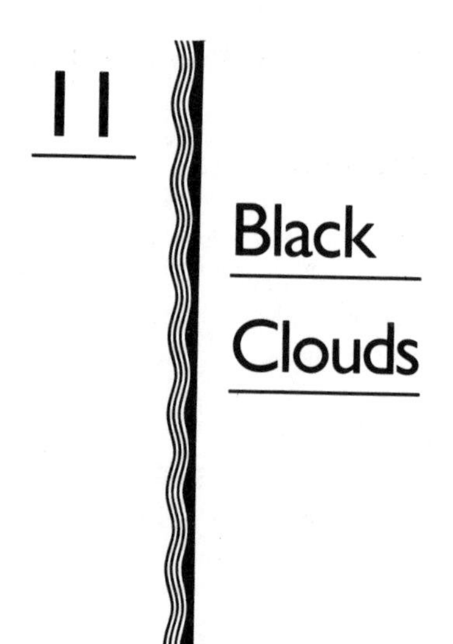

11 Black Clouds

One cloud is sufficient to eclipse a whole sun.
Baltasar Gracián: *Oraculo Manual y Arte de Prudencia*

Several years ago, when I worked at the University of Arizona, I ended each day by marching off to slake my thirst with one of my colleagues, Sean Twomey. As we would leave our building, there often would be small, scattered clouds in the sky. One of us would point to the darkest of them and say, "Look at that black cloud, it must have a lot of carbon in it." This would be followed by raucous laughter. It was a private joke, and repeating it got to be a kind of ritual that never palled.

We were able to share this joke because both of us had heard people solemnly ascribe the blackness of clouds to carbon or soot in them. These people were not unlettered peasants but rather those who bloody well ought to have known better, which was why their pronouncements were so amusing. It is the idiocy of savants that evokes derision, not that of those unprocessed by some pedagogical slaughterhouse.

Although we usually think of clouds as being white, perhaps the whitest objects of our experience, some clouds are not very bright or at least less bright than their surroundings, hence we call them dark or even black. Soot, a catch-all for the carbonaceous particulate products of combustion, is black because carbon in most of its forms strongly absorbs visible light. But merely because an object is black does not infallibly signal the presence of some strongly absorbing substance. Clouds of absorbing particles may be black, but all black clouds are not composed of such particles. Indeed, whether we call a cloud black depends on how it is illuminated and what its surroundings are. Even how we view it determines whether it is black, which can be driven home by a simple demonstration.

EXTINCTION

Suppose that we put some particles (they could just as well be molecules) between a beam of light and a detector. Although it could be the human eye,

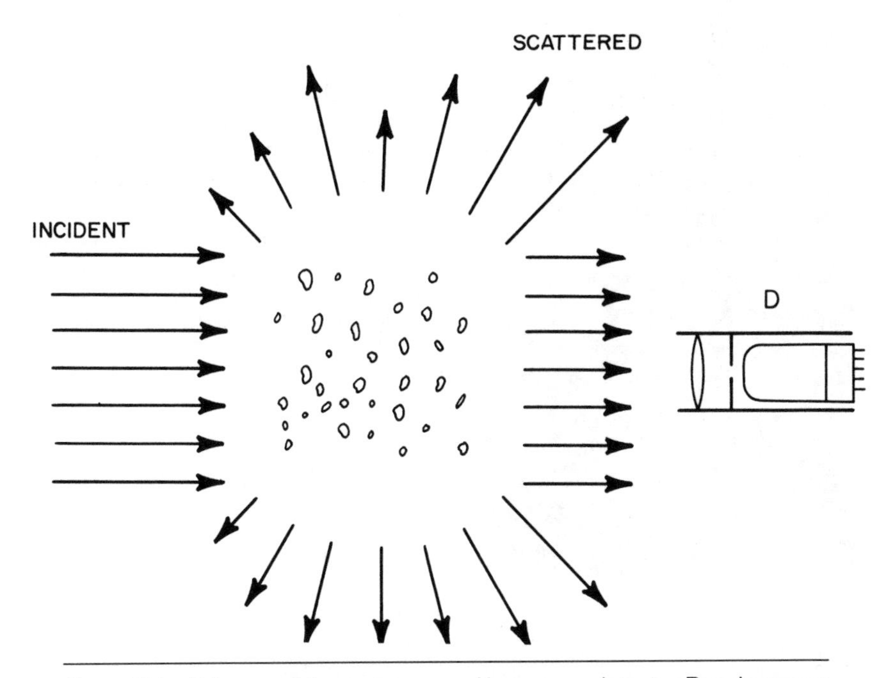

Figure 11.1 When particles are interposed between a detector D and a source of light, the detector receives less light than it would otherwise because the particles absorb some of the incident light and scatter some of it in all directions. This combination of scattering and absorption is called extinction. From *Absorption and Scattering of Light by Small Particles.* C.F. Bohren and D.R. Huffman (Wiley-Interscience, New York, 1983).

the detector shown in Figure 11.1 is an electronic detector, which like the light source is collimated (i.e., confined to a narrow set of directions). The detector's response is proportional to the amount of radiant energy it receives, which *decreases* when the particles are in place. The incident beam is said to have been extinguished, or attenuated, or to have undergone extinction. That is, less light is transmitted to the detector. Some of the light from the incident beam has been scattered by the particles and some of it has been absorbed by them. Absorption irretrievably removes light from a beam by transforming it into other forms of energy, whereas scattering merely redistributes it into other directions. By measuring only extinction, however, we cannot determine how much of it is apportioned between scattering and absorption; we can measure only the *sum* of the two. A simple demonstration of this rather elementary—but sometimes overlooked—point was devised by another of my Arizona colleagues, Donald Huffman.

Take two Petri dishes (or some other transparent dishes) filled with clean water and place them on an overhead projector. Their images projected onto a screen will be identical. To one dish add some milk; to the other some India ink. The projected images of these dishes will become dark, but still indistinguishable

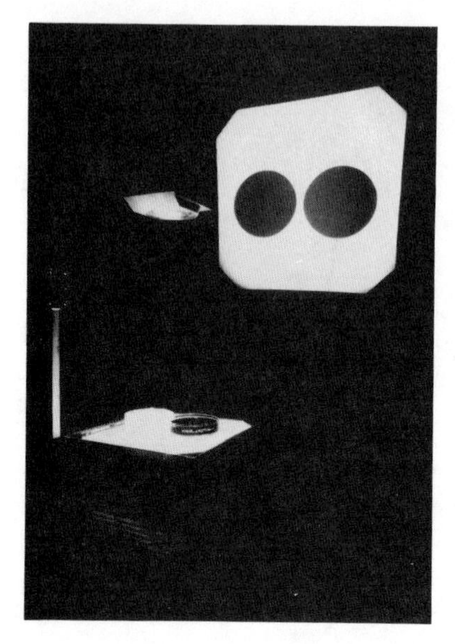

Figure 11.2 The images of the two water-filled Petri dishes on the screen are identical, yet their darkness is caused by different mechanisms. Light incident on the inky water is not transmitted to the screen mostly because of absorption, whereas light incident on the milky water is not transmitted mostly because of scattering. It is only when we look at the dishes rather than at the screen that this difference becomes evident.

(Fig. 11.2). Both Petri dishes are dark when viewed by the light they transmit. Merely by looking at the images on the screen, it is impossible to tell which originates from the dish filled with milky water and which from the dish filled with inky water. What we observe on the screen is a consequence of extinction, which to apportion between scattering and absorption requires a second, independent, observation. For example, if we look directly at the Petri dishes, it is obvious which one contains the milky water and which the inky water.

I wish that I had two photographs of the same cloud, one taken from above it, the other from below it. The best I can come up with is two photographs of different clouds taken under different circumstances. They are shown in Figure 11.3. The top photograph was taken in State College. It shows a dark, ominous cloud over Mt. Nittany, a mountain held sacred by the natives there. This cloud strongly attenuates the much brighter light behind it. The bottom photograph of a cloud-filled valley in South Wales shows clouds brighter than their surroundings. These are seen by reflected light. Had I been under them, I would have seen the less bright light that had *not* been reflected (i.e., the transmitted light).

A WARNING

I am fond of saying that the universe was not created for the convenience of those who frame multiple-choice examinations. Lest I be guilty of the same misdemeanors against which I inveigh, I must point out that my discussion of extinction in the preceding section goes only part way toward explaining why some clouds may appear black. Suppose that you are in an airplane that is taking off on a gloomy day. The sky overhead is filled with clouds. Thus it is darker

Figure 11.3 The clouds in the top photograph are darker than the background skylight, which is partly extinguished (mostly by scattering) by them. The clouds in the bottom photograph, seen by reflected rather than transmitted light, are brighter than their surroundings.

at ground level than it would be otherwise because of extinction: only a fraction of the sunlight incident on the clouds is transmitted by them. From observations on the ground it is not possible to say whether this extinction is predominantly scattering or absorption. It is only when your airplane climbs above the clouds and you see how bright they are that it is now evident that extinction of visible light by them is mostly scattering: incident sunlight was not irretrievably lost, it was merely redirected. This is an example to which the demonstration strictly applies. There are instances, however, in which it sheds little light on what is observed. I shall have more to say about this in Chapters 14 and 18.

Depending on how you do the demonstration discussed in this chapter, you may notice something that I did not mention. If you put just a tiny amount of milk or ink in the water, the transmitted images are reddish or reddish-brown. This is because extinction by milk and ink is *selective:* it depends on the wavelength of the incident light. You won't notice this with lots of milk or ink because extinction of visible light of all wavelengths by them is so great that the projected images of the Petri dishes are dark. Cloud droplets scatter visible light of all wavelengths about equally, so for my purposes it was not necessary to discuss selective extinction. At this point I could do so, but it gives rise to such interesting phenomena that it deserves its own chapter.

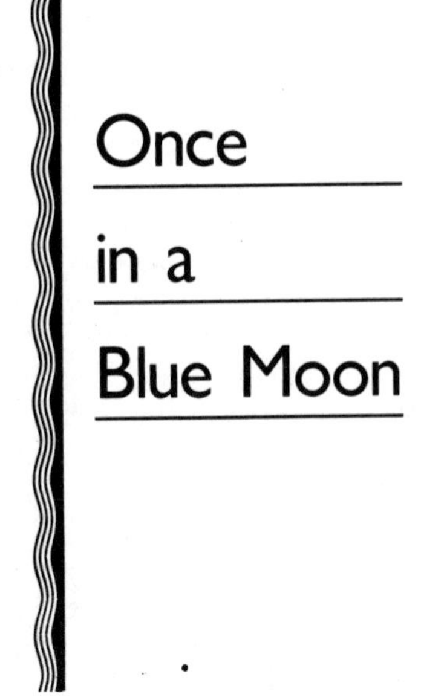

12 Once in a Blue Moon

Yf they saye the mone is belewe
We must beleve that it is true
Anonymous—16th century
Charles E. Funk: *A Hog on Ice*

Rare events are sometimes said to occur only "once in a blue moon." This saying is familiar enough, but how many of us have ever thought about its meaning? Is it merely whimsical, or is it a reflection of folk wisdom which has a basis in fact?

We usually see the moon as white or perhaps yellow. Low in the sky it may appear orange or red. It never is blue. Never? Well, never in my experience, but I have not been so lucky as Robert Wilson, who in September of 1950 observed a blue moon (and a blue sun) in Edinburgh. Of course, these unusual events were witnessed by many other residents of this city. But Wilson had a unique opportunity to do more than just note the curious fact that the sun was blue and then go about his business: he was an astronomer on the staff of the Royal Observatory (now professor of astronomy in University College, London) and had the presence of mind to make quantitative observations with a telescope. Moreover, on the basis of his and other observations and some calculations, he concluded that the blue sun and moon were caused by clouds of small particles from forest fires in Alberta, which had been carried by winds across the Atlantic to Europe. These particles were probably small oil droplets formed in the combustion products of the fire. Oil, unlike water, has a sufficiently low vapor pressure that small droplets of it can survive a trip across the Atlantic.

To my knowledge a blue moon has not been reported since 1950. So I must admit with some sadness that it is unlikely that I shall ever see one. But blue moons, however rarely they may occur naturally, can be produced artificially at will with the simplest of apparatus. This was happened upon by Donald

Huffman, a physics professor at the University of Arizona, during a lecture to astronomers. His topic was interstellar dust, the particles responsible for, among other things, the dark patches in the Milky Way. Dust between us and a star reddens its light: the starlight we receive is redder than it would be if there were no dust (this is not to be confused with the red shift, which is an entirely different matter). A familiar example is the sun, the star nearest to earth, which may be red at sunrise and sunset because of selective scattering by atmospheric molecules and particles.

Huffman usually illustrates his lectures with simple demonstrations. To demonstrate reddening of starlight he blew a puff of cigarette smoke into a beaker inverted on the glass plate of an ordinary overhead projector. The image of a clear beaker transmitted onto a screen is white; that of a smoke-filled beaker is reddish. At least that is what it is supposed to happen. And so it did—at first. But to Huffman's surprise the image of the smoke-filled beaker began to change from red to blue. And all the while he was explaining to an increasingly skeptical audience that starlight is reddened by particles such as the ones in the beaker. I don't remember how he managed to extricate himself from this embarrassing situation. But I do remember that after his lecture he came to my office in a highly excited state and insisted that I look at something unusual. We got our hands on an overhead projector, some beakers, and cigarettes, and with a bit of experimenting we convinced ourselves that what we saw on the screen was real, neither an artifact of the projector nor a hallucination. With a bit of practice Huffman was able to blow either red or blue smoke at will—both with the same cigarette, beaker, and projector. Before I tell you how to do this yourself, it is well to discuss briefly some of the underlying physics. I begin with familiar sights as stepping stones to more exotic ones.

SELECTIVE EXTINCTION

The sun is a source of white light: it emits visible light of all wavelengths in nearly equal proportions. Were it not for the atmosphere, sunlight would be transmitted to us without change in color or brightness. But the atmosphere contains molecules as well as highly variable amounts and kinds of liquid and solid particles. Let us assume that they scatter visible light without appreciably absorbing it, in which instance extinction is dominated by scattering (see the previous chapter). If visible light were scattered uniformly regardless of its wavelength, sunlight would be diminished in brightness without change in color as it traversed the atmosphere. Depending on their size, however, atmospheric particles (and molecules) scatter light of some wavelengths more than others. For example, if the scatterers are small compared with the wavelengths of visible light, they scatter blue light more than red. Such particles lying between us and a source of white light act as a filter which selectively removes more blue light from the beam than red. The more particles, the more the transmitted light is filtered. Consequently, the most brilliant reds are at sunset because it is then that the

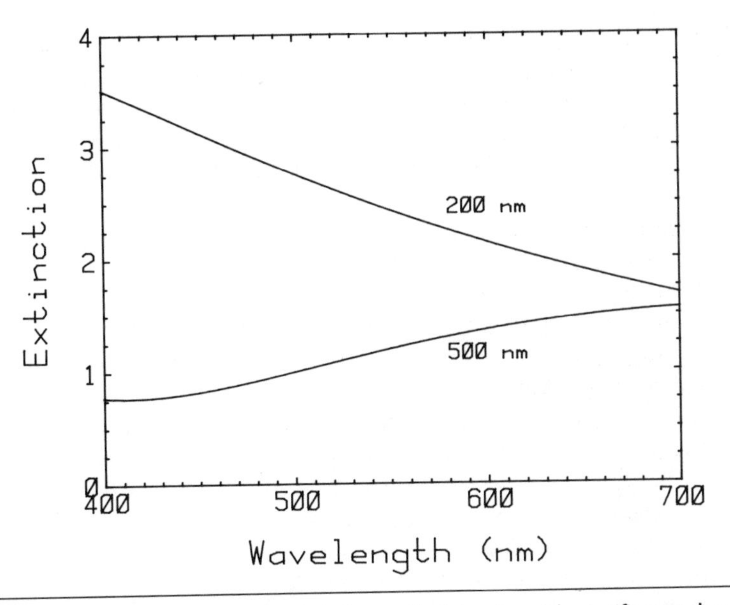

Figure 12.1 Extinction of visible light by small oil droplets. No significance should be attached to the absolute values of extinction shown, only relative values.

rays of the sun traverse the longest atmospheric paths. Regardless of the sun's elevation, its light is usually redder—relatively depleted of shorter wavelengths—than it would be if there were no atmosphere, even one free of all particles. Note, however, that for this to happen the scatterers must be smaller than the wavelength.

But it is not a universal law that only such scatterers may inhabit the atmosphere. Cloud droplets are much larger than the wavelengths of visible light. They scatter visible light of all wavelengths about equally. What about intermediate particles, neither much larger nor much smaller than the wavelength? For sizes comparable with the wavelength the reverse of what we have come to expect as normal can occur: red light is scattered more than blue (see Fig. 12.1). Particles of the proper size seem to be generated rarely by natural processes, but when they are, they can cause bluing of moonlight and sunlight rather than reddening. This is what happened in September of 1950: the particles carried from the Canadian forest fires to Scotland were just the right size to scatter red light more than blue.

SMALL-SCALE BLUE MOONS

An inverted beaker filled with cigarette smoke is a miniature polluted atmosphere; a projector lamp may be likened to the moon. But there the similarity ends because the blue moons that are so rare on a large scale can be made quite readily on a small scale: it all depends on how long you hold the smoke in your mouth before filling the beaker. The light transmitted by the smoke is likely to be reddish if you release it immediately, but if you pause for a moment the light may be blue. The holding time required to give blue images on the screen may be so short that onlookers will be baffled. Sometimes they do not notice any differences in how blue and red light are produced, even after you let them in on the secret. This is particularly true in a dry environment; conversely, the greater the relative humidity the more difficult it is to control the outcome. But with practice you should be able to create blue and red moons alternately like a sorcerer. You might want to experiment with different beakers and cigarettes; and do not overfill the beaker or little light will be transmitted.

I stated previously that blue light is scattered more than red by sufficiently small particles; the converse can occur if they are larger. It seems obvious that by holding smoke in your mouth for a moment the particles somehow become larger than they would be otherwise. The mechanism for this which first comes to mind is condensation of water vapor onto the dry particles, especially in light of Chapters 2 and 7: they either acquire a coating of water or, if soluble, dissolve in the condensed water. In both instances they are larger. Unfortunately, this explanation does not stand up under scrutiny. I have calculated that the time it takes for particles to grow to the size where they extinguish red light more than blue is so short that it would be impossible for you to get the smoke out of your mouth before they had grown too large. Also, if growth by condensation is indeed the mechanism it should be possible to transform a miniature blue moon to a red one by driving the water from the particles by heating. My attempts to do so gave inconclusive results: the color did not revert but perhaps I did not try hard enough.

Particles can also grow by coagulation: when they collide they stick together. In a dry, confined space smoke particles do not seem to grow to a size sufficient to transmit blue light preferentially. But they do so in a moist, confined environment. So perhaps the greater amount of water vapor in the moist environment merely serves to induce particles to grow to a larger size by coagulation.

That miniature blue and red moons can be made at will is indisputable. But precisely how they are made is still an open question.

A SPURIOUS BLUE MOON

Absorption by the particles considered in this chapter is negligible; it is not zero (no such particles exist), merely of little consequence to us here. Extinction therefore results mostly from scattering. Yet it is useful to distinguish between extinction and scattering, even when absorption is negligible. This is why I labeled the vertical axis in Figure 12.1 as extinction rather than as scattering, even though either would have been correct.

To illustrate the difference between what is conveyed by the terms extinction and scattering I note an observation that I once happened to make. On a partly overcast day I noticed a long column of smoke being spewed from a large stack on a rooftop. The lower end of the column was bluish, but well above the stack opening the smoke abruptly turned reddish. With the cigarette smoke demonstration in mind, one might be tempted to conclude that this was an example of particles shrinking as they ascended accompanied by a change in their extinction. This explanation should become suspect when I give you more details of what I observed. The bluish smoke was seen against distant dark ridges, and just where the background changed from ridges to bright clouds the smoke became red. Was this merely a coincidence, or was it a clue?

When the smoke's background is dark, one sees it mostly by light that *has* been scattered. When the smoke's background is bright and white, however, one sees mostly light that *has not* been scattered, that is, the transmitted light. If the scattered light is blue, that transmitted (i.e., that which has not undergone extinction) is reddish. You can observe this with the smoke-filled beakers of the demonstration. When the transmitted light is reddish, the scattered light—which you can observe by looking directly at the beaker—is bluish. And when the light transmitted is bluish, that scattered is reddish.

If absorption is negligible, extinction is equal to scattering, although these two terms may be used to convey different meanings. Extinction signals that transmitted light is of interest whereas scattering signals that scattered light is of interest.

COLOR AND WAVELENGTH ARE NOT SYNONYMOUS

If you have stared long and hard at Figure 12.1 while weighing my arguments in the preceding paragraphs, you may have a few doubts nagging you. For example, scattering by the 200 nm particles is greatest not for blue light but for violet light. If we look at a cloud of such particles by scattered light, shouldn't it therefore appear violet? And suppose we look at light transmitted by a cloud of the 500 nm particles? Extinction by such particles is least in the violet, so shouldn't the transmitted light appear violet?

In Chapter 10 I discussed the emission of radiation, mostly infrared radiation. I purposely did not equate wavelength with color, as writers of popular books and textbooks often do. I suppose they hope that their readers will grasp the perhaps unfamiliar concept of wavelength by way of something more familiar. Yet this is helping them over a curb while abandoning them at the base of a cliff. In the first place, it is meaningless to speak of the "color" of infrared radiation because we cannot see such radiation. And color has meaning only to the human observer (some animals and insects perceive color, but since we cannot communicate with them we have no way of really knowing if our perception of color is the same as theirs). To demote color to a synonym for wavelength is to sever its intimate relationship with the human observer.

Even if we restrict ourselves to visible light, it still is incorrect to equate color and wavelength. Yellow light is often taken to lie between 560 and 590 nm. Yet these limits are not absolute. Is light of wavelength 590.1 nm not yellow? This question is meaningless: the human observer is incapable of distinguishing between 590 and 590.1 nm light. Further evidence that the color yellow is not uniquely specified by a particular wavelength or range of wavelengths is that by mixing green and red light we can obtain yellow light that gives the same visual sensation as light of wavelength 590 nm. Even if we ignore the fuzzy boundaries between the somewhat arbitrary wavelength intervals defining various colors, we are still faced with the undeniable fact that there is no unique relationship between a perceived color and the spectrum of light that evokes it.

Now we are better equipped to understand why sunlight scattered by 200 nm (or smaller) particles is blue rather than violet. When you look at any source of light, the sky for example, you do not perceive the relative amount of light in each wavelength interval. You call the sky blue, and can say no more. But skylight is not blue in an objective sense; it is not blue light and nothing else, even if we ignore the ambiguity of the term "blue." Skylight is light of all visible wavelengths, blue merely predominates, not solely because of the wavelength dependence of scattering by atmospheric molecules (greatest for violet light) but also because of the wavelength dependence of the response of the human eye. Although violet light is indeed scattered more than blue light by very small particles, the human eye is less sensitive to violet than to blue. As a consequence, we perceive the light scattered by such particles as blue, not violet.

In explaining the blue moon demonstration and the chimney smoke observation I said nothing about scattering of violet light. And I shall not say much about it in the following chapters. This does not mean, however, that we cannot see violet light in nature. The rainbow (see Chapter 21) exhibits violet. But note the difference between violet light in the rainbow and in the sky. Violet light in the rainbow is separated from light of other colors whereas in skylight it is not.

The statement that blue light is scattered more than red light is usually merely shorthand for saying that violet light is objectively scattered even more than blue light, but because of the response of the eye, the human observer subjectively perceives blue light to be scattered more than violet.

SET AN ASTRONOMER TO CATCH AN ASTRONOMER: A POSTSCRIPT

Not long ago several of my students gleefully brought me articles clipped from the yellow press. It seems that an astronomer had gotten the ear of credulous reporters to whom he asserted that a blue moon has nothing do with its color, it is merely what astronomers call the second full moon in the same month. Although I am inured to the sometimes quaint and curious terms used by astronomers, it does seem a bit silly for one of them to deny the objective existence of the blue moon after another one has measured its spectrum.

13 The Green Flash

". . . do you ever get to see that flash of green? People say they've seen it, the minute the sun disappears."

"Now it stands to reason, mister, any damn fool stares into the sun long enough, he'll end up seeing exactly what some other damn fool tells him he's going to see."
John D. MacDonald: *A Flash of Green*

It does not take much wandering around the coastal cities of Southern California before coming to the conclusion that every man and his dog has heard of the green flash. But no one sober enough to be believed admits to having seen one. Perhaps this explains the high divorce rate: there is a Scottish legend that anyone who has seen the green flash will never err in matters of love. Yet the West Coast of the United States is an ideal place for observing the green flash, which is not a rare phenomenon. It is indeed true that the green flash is rarely observed, but not because it is rarely observable. The sad truth is that most people are only dimly aware of their surroundings. Perceptive observers are rare, not green flashes. One of my colleagues lives on a farm nestled among the rolling hills of central Pennsylvania. To him the green flash is commonplace. He sees it frequently because he knows what to look for and is willing to look.

At the very least, the *green rim* of the sun, an essential ingredient in the production of a green flash, may be seen with binoculars as the sun sets over the Pacific on any reasonably clear evening. Indeed, it may be seen as the sun sets—or rises, for that matter—in many parts of the world, not just coastal regions.

It was contemplation of the fate of all those Southern Californians who will go to their graves without having seen a green flash—a tragic example of poverty in the midst of plenty—that inspired the following demonstration.

A GREEN RIM DEMONSTRATION

An opaque slide with a hole in it together with a slide projector provide the means for simulating the sun. The projector is a source of white light, a mixture of all colors, and the cross section of the beam is circular. As a consequence, a white disk is seen when the beam is projected onto a screen. We may consider this disk to be made up of many disks with all the colors of the visible spectrum, but because they coincide the net effect is a white disk. If the light is forced to follow a more circuitous path, however, the disk changes.

Place a mirror at the bottom of a shallow pan filled with clean water (the mirror will have to be propped up a bit) and direct the projector onto this mirror at an oblique angle to the surface of the water, the greater the angle the better (Fig. 13.1). What you will now see on the ceiling is a predominantly white disk except that one of its rims is red and the other violet. When the projector beam encounters the boundary between the water and air it changes direction, that is, it is *refracted*. Because all the separate components of the beam—red, orange, yellow, green, blue, violet—are refracted differently, the disks do not quite coincide. In the region where they do the disk is white, where they do not we see colors; this is because violet light is refracted most and red light least. Note that the disk is no longer circular; refraction in this instance not only causes color separation, it also distorts the circular disk into an elliptical one.

Milk is a suspension of particles that are small compared with the wavelengths of visible light. Consequently, these particles scatter light of shortest wavelengths most (see the previous chapter). Thus if you add a *small* amount of milk to the

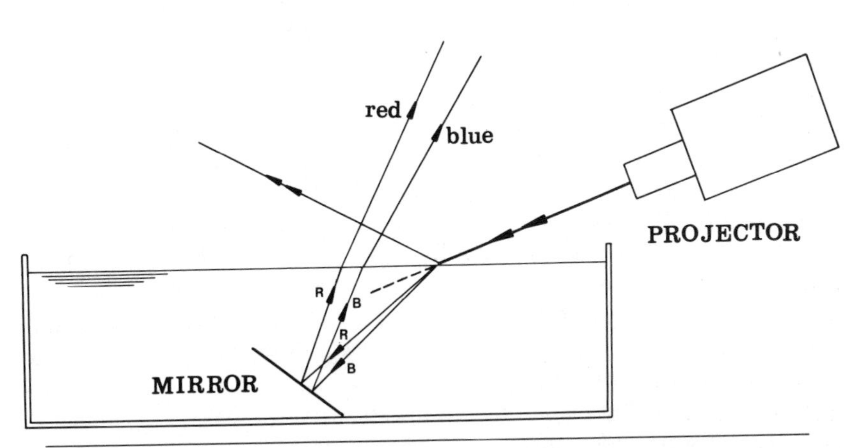

Figure 13.1 Because light of different colors is refracted different amounts, blue more than red, white light from a projector is separated into its components when the beam is directed onto a mirror at the bottom of a shallow pan filled with water. For clarity, the separation of the blue and red components is exaggerated.

water in the pan shown in Figure 13.1, the beam illuminating the mirror and that reflected by it will be relatively depleted of their shorter wavelength components. As more milk is added the disk becomes less intense because light has been scattered out of the beams. And the light that does get to the ceiling is richer in longer wavelengths because the light of shorter wavelengths has been preferentially removed. With enough milk in the water the central part of the disk is no longer white. More important, its rim is no longer violet or even blue—it is green. It is the combination of preferential refraction and scattering that gives the disk its green rim. This is sufficient to explain, at an elementary level, what is observed. Yet it may not be completely satisfactory, and some of you still may be nagged by a few questions. For example, if refraction and scattering occur when light interacts with matter, what is the difference, if any, between these two processes? Before connecting the demonstration with what is observed in the atmosphere, I address this question.

REFRACTION AND SCATTERING

What are called the "laws" of reflection and refraction were derived from observations made long before the nature of matter and light was well understood. The law of reflection—angle of incidence equals angle of reflection—is the more ancient of the two, having been clearly stated by Hero of Alexandria in the second century A.D. And although the law of refraction (Snel's law) emerged from comparatively recent investigations, it is still nearly 400 years old. In elementary textbooks these laws are presented in ways that neither Snel (this is the way *he* spelled his name, so don't reach for your blue pencil) nor Hero would find startling. Yet physics has taken a step or two forward since Snel's day, and it seems reasonable that a fuller understanding of optical laws might be acquired by appealing to somewhat newer findings.

There are two ways of thinking and talking about light: (1) as *waves* or (2) as particles, called *photons*. Much ado is sometimes made about this duality, although needlessly, in my opinion. Duality is not necessarily a defect in our way of thinking about light: Who would be so hidebound as to insist that Nature can be comprehended only by rigidly adhering to a single interpretation? Sometimes it is more appropriate to adopt the wave interpretation whereas at other times the photon interpretation seems more natural. The choice is dictated solely by simplicity: that which is simplest is best. Only fools choose to make something more complicated than it need be. Sometimes it is even appropriate to switch interpretations in mid-explanation; I shall do so shamelessly in the following paragraphs.

After a photon is born (i.e., emitted; see Chapter 10) it can suffer only one of two possible fates when it interacts with matter: (1) it is *absorbed,* that is, it ceases to exist, although its energy is taken up by whatever it interacts with; or (2) it is *scattered,* in which instance it survives the interaction intact but possibly changes direction (see the previous two chapters for more on this). In the present context absorption is negligible, so the photons with which we are concerned are only scattered. If you accept this you must also accept that refraction is ultimately a scattering phenomenon. What, then, should we make of the statement that refraction is "bending" of light? It is a metaphor: a beam of light is not bent in the same way that, say, an iron bar is bent. Yet when a beam of light is incident on the smooth interface between two dissimilar substances, what is observed is described most economically and simply by saying that the beam is bent by an amount given by the law of refraction. We therefore agree to call something a refraction phenomenon if it can be explained satisfactorily by appealing to an *approximate* relation called the law of refraction. But it would be absurd to go beyond this and assert that refraction (or reflection) and scattering are fundamentally different processes. Refracted photons do not bear scars of their interaction distinguishing them from scattered photons; refracted photons and scattered photons do not have different stories to tell about their experiences: a refracted photon *is* a scattered photon. Yet in seeming contradiction to this I stated at the end of the preceding section that the green rim of the disk owes its existence to refraction *and* scattering. To elucidate further the connection between refraction and scattering it helps if we now switch our allegiance to the wave interpretation of light.

In Snel's day it may have been suspected by some that matter is not a structureless continuum but is composed of discrete, indivisible units. Yet it was not until comparatively recently that sufficient evidence had been amassed to compel acceptance of the concept that matter is made up of atoms. And we also know that these atoms are composed of charged particles, electrons and protons. At about the same time that the existence of atoms was becoming increasingly difficult to refute it was discovered that light can be considered to be an electromagnetic wave. The electric field of such a wave can therefore exert forces on the elementary charges in matter and set them into oscillatory motion with the same frequency as that of the field driving them, hence they radiate, the reason for which was discussed in Chapter 10. Light scattered by atoms or molecules or particles is therefore excited electromagnetic radiation. Note the distinction between radiation excited and that emitted. All objects (at temperatures above absolute zero) emit electromagnetic radiation whether they are exposed to an external source of radiation or not. But to scatter radiation, matter must be excited by such a source.

Consider now a beam of light incident on an air-water interface (Fig. 13.1). The water is composed of a vast number of molecules, each of which is excited by the incident beam and therefore contributes to what is observed, a *superposition* of very many scattered waves and the incident wave. In the water the incident wave may be considered to have been extinguished because of interference with all the waves scattered by the molecules and replaced by another wave with a different direction of propagation. Scattering in this special direction is said to be *coherent*. The word "cohere" means to stick together, especially entities that are similar, and this is just what all the waves scattered by the individual water molecules do: in the direction of refraction the crests and troughs of all the waves scattered by the molecules very nearly coincide. As a consequence, total scattering in this special direction is quite strong. Not more than a century ago it was believed that if all foreign particulate matter were removed from a homogeneous substance (e.g., water or air), there would be no scattering in directions lateral to a beam traversing it. But we know better now: a beam of light in the purest of substances is *incoherently* scattered, although weakly, in lateral directions. Water is such a substance, and when milk is added to it the amount of incoherent scattering is greatly increased. So the green rim of the disk is caused by coherent scattering (refraction), whereby violet and blue light is deviated more than red, as well as incoherent scattering, which depletes the refracted beam of its shorter wavelength components. Because the human mind usually grasps a complex process more readily if it is imagined to be made up of simpler processes each of which is understood, the green rim is attributed to refraction and scattering. Perhaps refraction and *incoherent* scattering would be a better way of phrasing it.

THE GREEN RIM OF THE SUN

Refraction occurs when a beam of light propagating through some medium encounters a different medium. Yet the transition from one to the other need not be as abrupt as that shown in Figure 13.1. Except possibly in the bottom few meters or so the density of the atmosphere generally increases with decreasing height. As a consequence, rays of light entering the atmosphere are refracted, although their direction changes gradually rather than abruptly (Fig. 13.2). Moreover, atmospheric refraction of blue light is greater than that of red light. Although the amount of refraction is much less than that at an air-water interface, it still leads to observable consequences, especially when the sun is very low in the sky. Consider, for example, the rays of white light shown in Figure 13.2. Outside the atmosphere the blue and red components of these rays coincide, but as they progress downward through the atmosphere the blue component is refracted more than the red. An observer at O therefore sees the blue image of the sun displaced slightly upward from that of the red image, which makes the upper rim of the sun blue when it is near the horizon and the lower rim red. This is just like the disk in the demonstration, although the angular width of the blue rim of the disk is much greater than that of the blue rim of the sun.

When the sun is low in the sky its upper rim is blue only in the absence of appreciable scattering. At high elevations where the air is relatively cleaner, the sun's upper rim may be seen to be blue, although even at sea level this has been observed (see *Weather,* February 1972, p. 91). More usually it is green, just like the green rim of the disk, although not nearly as wide; this is the result of selective depletion of the shorter wavelength components of sunlight because of scattering by atmospheric molecules and particles.

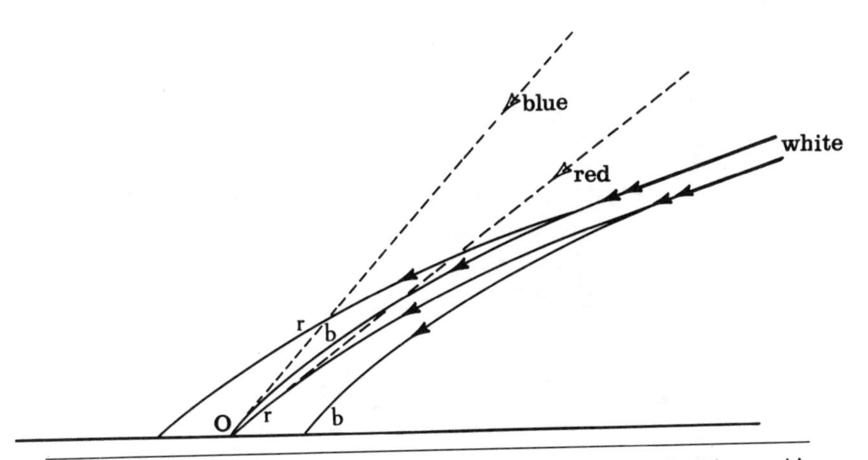

Figure 13.2 Refraction of sunlight (shown greatly exaggerated) in the earth's atmosphere. The red and blue components of white light are refracted different amounts.

I well remember the first time I watched through a telescope the sun set over the Pacific. I was trembling with excitement. My heart was pounding and I was dripping sweat. My excitement reached its peak when I could make out the green rim of the sun during the last few degrees of its arc. Under normal conditions the angular width of this green rim is too small to be resolved without the aid of binoculars or a telescope. A green flash, however, is seen with the unaided eye. If temperature gradients, hence density gradients, near the earth's surface are sufficiently large, atmospheric refraction magnifies the green rim momentarily as the sun sets, and a flash of green light may be seen. This requires special, although by no means rare, conditions. But the green rim of the sun, which is inseparable from the green flash, may be seen when the sun is near a low horizon in a cloudless sky. You don't have to be in Tahiti, although you might wish to be, which is another matter entirely. One of my colleagues at Dartmouth lives on top of a mountain in New Hampshire, as far away from Tahiti, both geographically and culturally, as one can get. Yet he sees the green rim of the sun every cloudless evening—except when the atmosphere is so clean that he sees its blue rim.

14

Multiple Scattering at the Breakfast Table

white . . . is not a mere absence of colour; it is a shining and affirmative thing, as fierce as red, as definite as black.
G. K. Chesterton

The scene shown in Figure 14.1 is familiar enough. In it are the kinds of things likely to be found on your breakfast table: salt, sugar, and a glass of milk. Although dissimilar in many ways, they all are white. Why this is so is the subject of this chapter. And I can think of no better way to begin than with a simple demonstration.

Figure 14.1 Multiple scattering at the breakfast table.

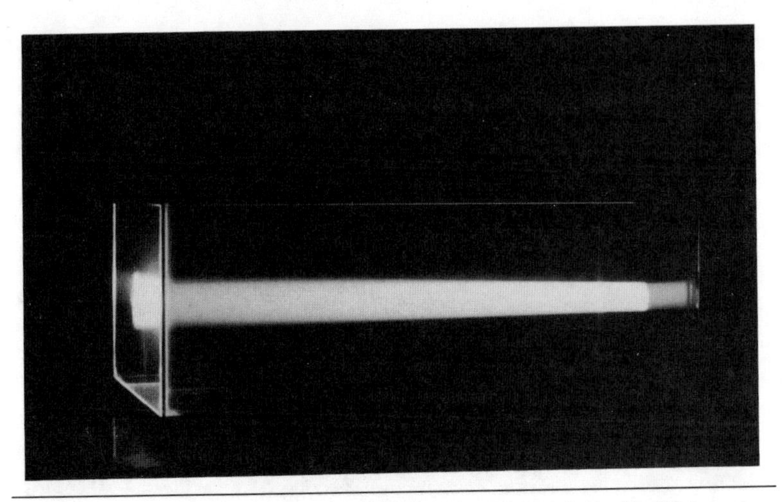

Figure 14.2 Single scattering by a suspension of fat globules (milk) in water.

SCATTERING IN AN AQUARIUM

Scattering of light by particles and by molecules was discussed in Chapters 11, 12, and 13. Without saying so, in these chapters I had in mind what is called *single* scattering in contradistinction to *multiple* scattering. The difference between the two kinds of scattering can be demonstrated using a slide projector, an aquarium filled with clean water, and some milk.

A slide projector in which an opaque slide with a hole in it is inserted provides a collimated beam of light. You cannot see this beam from the side unless it traverses a medium that scatters some of the light from the beam toward your eyes (this, by the way, is why movie scenes in which combating spaceships fire brilliant laser beams at one another are absurd: in space, where there is little to scatter light, you would see a beam only if it were pointed directly at you).

If the projector beam is shone into an aquarium filled with clean water you may, in a very dark room, barely be able to see the beam. To make it more evident add a few drops of milk to the water (Fig. 14.2). Milk contains tiny globules of fat which scatter light from the beam, and it is attenuated along the direction it propagates. Its margin is well defined as a consequence of single scattering: a photon (see the previous chapter) must be scattered at least once for us to see the beam (to be precise, we do not see the beam but rather the light removed from it), but its likelihood of being scattered more than once is small. To increase this likelihood add more milk to the water; that is, increase the concentration of scatterers. The result is shown in Figure 14.3. Note that light now comes from beyond the limits of the beam where previously none had been evident (Fig. 14.2). This is because photons scattered by particles in the beam are scattered again by particles outside it and thence to our eyes. Multiple scattering increases the number of ways in which a photon may reach our eyes.

Figure 14.3 Multiple scattering by a suspension of fat globules (milk) in water.

A PILE OF PLATES

Plates cut from clear plastic sheets provide raw material for demonstrating other aspects of multiple scattering. One such plate, on a black background, is weakly reflecting; but a pile of them is strongly reflecting (Fig. 14.4). Here is an example of an ensemble—a pile of transparent plates—with optical properties quite different from those of its individual members. As plates are successively added to the pile it reflects more light, but beyond a certain number, in which instance the pile is said to be *optically thick,* each additional plate yields an ever smaller increment. So it is also with any optically thick scattering medium, such as a cloud. A single water droplet does not scatter much light, but because of multiple scattering an optically thick cloud of such droplets reflects much of the visible light incident on it, which is obvious to anyone flying over a thick layer of clouds.

For a collection of scatterers (e.g., a pile of plates, a cloud) to be bright and white upon illumination by white light it must not only be optically thick but its members must only *weakly* absorb such light. Consider a pile of plates in photon language. Most of the photons incident on a single plate are transmitted to the underlying black surface where they are absorbed. But with two plates in a pile some of the photons transmitted by the first are reflected by the second. Each plate added to the pile increases the probability that a photon eventually gets to your eyes, so with enough plates the pile is white. Again, as with the aquarium demonstration, multiple reflection (scattering) increases the number of ways—reflected once, twice, and so on—in which a photon can get to your eyes. But multiple scattering takes as well as gives: the greater the number of scatterers the greater the chance that a photon will be absorbed; multiple

Figure 14.4 A pile of transparent plates on a black background.

scattering not only increases the pathways by which incident photons can re-emerge from a medium, it exposes them to a greater hazard of being absorbed. I shall return to this later.

WHITE CLOUDS: THE CONVENTIONAL WISDOM

It has been stated countless times that clouds are white because they are composed of droplets sufficiently large that they scatter visible light of all wavelengths about equally. Indeed, this notion is so widespread and has so many adherents that it now transcends science and has become an article of faith; to believe it is a mark of piety. I can therefore anticipate the great howling that will be raised in response to my assertion that this explanation is demonstrably false. At the very least, those who advance it fail to distinguish between a *necessary* condition and one that is merely *sufficient*.

For a cloud to be white (upon illumination by white light) it is sufficient that its droplets scatter visible light of all wavelengths about equally—but it is not necessary. If it were, a glass of milk would be blue. For milk is a suspension of particles that, unlike cloud droplets, scatter blue light more than red light; this was why milk was used for the green flash demonstration discussed in the previous chapter. Yet a glass of ordinary milk is white, not blue. Why? The single-scattering characteristics of particles in milk are really quite irrelevant to the appearance of a glass of milk. What is relevant is that it is optically thick—a small glass of milk looks no different from a large glass—and its particles are very weakly absorbing. Although incident photons corresponding to the color red may have to rattle around more in a glass of milk before re-emerging than those corresponding to blue, almost all of them ultimately re-emerge having escaped absorption. It is these two characteristics—optically thick and weakly absorbing—that unite the dissimilar objects with which I began this chapter. Regardless of their single-scattering characteristics, if enough weakly absorbing particles are heaped into a pile it will be white if the source of illumination is white. Besides the examples of milk, salt, and sugar in Figure 14.1 there is snow, flour, powdered glass, white sand—the list is endless. Taken individually these particles may have vastly different scattering properties; collectively, however, they are nearly identical.

Every glass of milk refutes the conventional explanation of why clouds are white. In fairness to those who espouse it, I might add that the scattering characteristics of individual particles in optically *thin* clouds are relevant to their appearance. For example, thin cirrus clouds are white—although not nearly as bright as cumulus clouds—because their particles are large compared with visible wavelengths; in contrast, noctilucent clouds—very tenuous high-altitude clouds—are often described as bluish, which is offered as evidence of the smallness of their particles. But I think it also fair to say that people who ask why clouds are white do not have in mind wispy cirrus clouds but, rather, proper clouds: big, fluffy, towering cumuli.

As a harsh critic of what I consider to be superficial explanations, I would be wise to forestall criticisms of mine. It might be—indeed, has been—asserted that reflection and scattering are fundamentally different processes, in which instance the two experiments described in preceding paragraphs are unrelated: the first demonstrates multiple scattering whereas the second demonstrates multiple reflection. Such logic-chopping would make a medieval theologian blush. I argued in the previous chapter that the term scattering embraces reflection as well as refraction. We continue to use these terms separately for historical reasons (i.e., intellectual inertia) and because they are convenient ways of summarizing what is observed. But other than metaphorically, light does not bounce off of a window pane. What we call the light reflected by glass is the sum of all the light scattered by its constituent molecules; if we ignore these details it is for convenience not by necessity.

CLOUDS AND SNOW

You may have noticed that clouds are often not as bright as snow (Fig. 14.5). This is a variation on the black cloud theme, which was discussed in Chapter 11. Of course, to compare fairly a cloud with a snowpack the conditions of illumination and viewing must be identical, and this is not always easy to obtain. Nevertheless, I have yet to see clouds brighter than the brightest snow. This is not because cloud droplets are more absorbing than ice grains in snow. Indeed, the reverse is true: the ratio of incident light scattered to that absorbed

Figure 14.5 All else being equal, snowpacks are usually brighter than clouds, as in this winter scene near Alta, Utah.

by a single cloud droplet is *greater* than that by single ice grains in snow because the grains are much larger (see the next chapter for more on this).

Both clouds and snowpacks are multiple-scattering media. The difference between them lies in their different *optical* thicknesses, that is, their thicknesses measured in units of mean free paths (see Chapter 16). Snowpacks are usually optically thicker than clouds. Indeed, snow on the ground is often effectively infinitely optically thick. Except for very shallow snowpacks, the addition of another layer does not sensibly change the fraction of the incident light it reflects. You may have observed this many times. Snow a few inches deep is indistinguishable from that a few feet deep. Clouds are darker than snow not because they absorb more light—they absorb less—but because they transmit more of it to their surroundings.

AN APPARENT PARADOX

Before discussing yet another aspect of multiple scattering, there is one other matter to be disposed of. It may have occurred to you that every cloud presents what at first sight is a paradox. Water molecules, like all the other molecular constituents of the atmosphere, scatter light. But when a given amount of water vapor condenses to form droplets the resulting cloud scatters much more than the water molecules did. This was evident in the cloud bottle demonstration of Chapter 2. It is not the amount of water in a cloud that makes it appear so different from the patch of sky in which it was born but rather the state of aggregation of this water. Why? The answer lies in the concept of coherence, which I discussed briefly in the previous chapter.

A group of water molecules when randomly separated scatter incoherently; when these same molecules condense into a droplet (i.e., they all are part of the same entity rather than independent agents) they scatter coherently. And coherent scattering can be much more intense, all else being equal, than incoherent scattering. Suppose that a group of hare-brained people is trying to push a stranded car. If they push incoherently (i.e., each individual pushes randomly when and in whatever direction he chooses) the car is not likely to do more than rock back and forth, possibly inching forward laboriously. Now suppose that a leader emerges from the group and urges everyone to push coherently: One, two, three, Push! In no time at all the car will be on its way. I have implicitly assumed that the pushing is done coherently *in phase:* coherence by itself is not sufficient unless directed to the same end (I shall have more to say about coherent scattering in Chapter 18). Just as coherent (in phase) pushing yields a much greater effect than incoherent pushing, so also is coherent scattering vastly more intense even though the scatterers are the same in both instances.

While I am on the subject of coherence and incoherence I should note that although a single cloud droplet is a coherent scatterer, a group of them scatters incoherently. If you are reeling after this apparent contradiction, I'll back up a bit. A group of N isolated water molecules in the atmosphere scatters incoherently: scattering by N is N times scattering by one. But when these same N molecules condense to form a droplet, scattering by them is much greater. Now consider a group of water droplets. If they are as far apart as those in atmospheric clouds, scattering by N droplets is N times scattering by one. That is, the droplets considered as a group are incoherent scatterers, whereas each droplet itself is a group of coherent scatterers. The multiple scattering I have discussed in this chapter (and will discuss further in subsequent chapters) is incoherent multiple scattering. Now let us imagine that a cloud of water droplets is compressed until it forms a continuous film. Again, what is observed is predominantly coherent scattering, more intense than incoherent scattering by the cloud but confined mostly to two directions. For example, sunlight reflected by a film of water is much brighter than that reflected by a cloud, at least a thousand times greater, possibly much more depending on the angle of incidence of the sunlight (to convince yourself of this compare the reflection of the sun in a pond with light from a cloud). But this reflected light is concentrated in a single direction, that given by the law of reflection, whereas the cloud scatters light in all directions. Light from the fogged mirror of Chapter 7 has about the same brightness in all directions; the unfogged mirror is much brighter, but only in one direction.

In going from a group of independent molecules to independent droplets to a water film, nothing changes except the arrangement of the molecules. Chemically they are the same. Yet what is observed is markedly different depending on the degree of coherence—the extent to which the molecules stick together, both figuratively and literally—of their scattering.

ANOTHER PILE OF PLATES

As long as we have plastic plates at hand we might as well use them to demonstrate yet another salient characteristic of multiple scattering: a little bit of absorption goes a long way. In the following section the relevance of this demonstration to the atmosphere is discussed.

I implied that the plates shown in Figure 14.4 are weakly absorbing. And so they are. But they are not nonabsorbing. To show this, take a single plate and lay it on a white background. Then add plates. With each additional plate the pile gets darker (Fig. 14.6). Multiple reflection exposes photons to multiple opportunities to be absorbed by the plates. The piles of plates in Figures 14.4 and 14.6 do not look the same because of different contrast and because they were photographed under different illumination and at different exposure times. Yet the two piles have the same reflectance (fraction of incident light reflected), which is independent of the underlying surface if the pile is optically thick. You can verify this with a pile of plates that straddles the boundary between black and white pieces of paper.

Figure 14.6 The same pile of plates as in Figure 14.4 but on a white background.

CLOUDS AT INFRARED WAVELENGTHS

The clouds we perceive to be so bright are much darker at infrared wavelengths. This is because absorption by water—liquid or solid—increases at wavelengths longer (and shorter) than those of visible light. And because of multiple scattering, absorption does not have to be great before a cloud of water droplets becomes quite dark; most respectable clouds will be black to radiation with wavelengths greater than 3 μm (the wavelengths of visible light lie between about 0.4 and 0.7 μm). We cannot see such dark clouds, of course. What we do see often misleads us.

One sometimes encounters—on examination papers if nowhere else—the assertion that it is warmer on cloudy nights because the clouds "reflect" infrared radiation back to earth. They do no such thing, no more than the tree and the sand in Chapter 10 reflect it. Water is so much more absorbing of infrared radiation emitted by terrestrial objects than of visible radiation that all but the thinnest clouds absorb nearly all the terrestrial radiation incident on them. Clouds also emit infrared radiation, which is where the confusion lies. Such radiation is not derived, except indirectly, from the incident terrestrial radiation; in contrast, the light we see from clouds has been derived from incident sunlight: we do not see them at night. Clouds, however, emit day and night; they keep the earth warm at night because they emit more infrared radiation than the clear sky.

MULTIPLE SCATTERING AT THE BEACH

Multiple scattering is such an intersting topic and there are so many examples of it in nature that I shall devote the next chapter to exploring it further. I invite you to leave the breakfast table and go to the beach.

15

Multiple Scattering at the Beach

The wind is rising on the sea
The windy white foam-dancers leap
Arthur Symons

In the previous chapter I discussed multiple scattering, taking as my theme objects commonly found on breakfast tables. At the end of that chapter I promised an excursion to the beach. What we might see there is shown in Figure 15.1. The sand is dark where it is lapped by waves, but farther up the beach

Figure 15.1 This beach on Point Loma provides many examples of multiple scattering. Note that the dry sand is lighter than the wet sand.

the sand is light. And note the foam sparkling on the breaking waves. All this is a consequence of multiple scattering.

Recall that a multiple-scattering medium is one containing enough scatterers that photons traversing it are likely to be scattered more than once. Strict single scattering would require a universe devoid of everything except a solitary scatterer and a beam of light. So the designations single and multiple are relative rather than absolute. Photons have a high probability of being scattered more than once in a multiple-scattering medium whereas this probability is low—but not zero—in a single-scattering medium.

For brevity I shall call any multiple-scattering medium simply an object. And throughout this chapter I shall always have in mind the brightness of objects as viewed by reflected light; that is, as viewed in directions opposite the source of illumination, which is sunlight if we are at the beach.

Light reflected by objects is multiply scattered light; it is the cumulative effect of many scattering events. Photons that reach our eyes have therefore suffered different fates: scattered once, twice, and so on. Of course, photons scattered many times are indistinguishable from those scattered only once: we perceive only images of varying brightness. For example, in Figure 15.1 the dry sand is brighter than the wet sand; the foam is brighter than both. You probably have no difficulty accepting that foam is brighter than sand, wet or dry; after all, bubbles are not at all like sand grains. But why is wet sand so much darker than dry sand? It surely cannot have anything to do with absorption by water, which is quite transparent to visible light. At least pure water is transparent, perhaps sea water contains enough impurities to darken the sand. The easiest way to check this is by performing a simple experiment.

DRY SAND AND WET SAND

Put some clean aquarium sand in a dish and wet half of it with distilled water. The result is shown in the left side of Figure 15.2. Note that the wet sand is darker than the dry sand, just as it is on the beach in Figure 15.1. To dispel the notion that the sand may contain impurities that are dissolved by the water, thoroughly wash the sand and dry it before performing this experiment; the results will be similar. So it is safe to conclude that impurities in water have nothing to do with why wet sand is darker than dry sand.

It is instructive to do this experiment using different liquids. For example, in the right side of Figure 15.2 half of the sand is wetted with water, the other half with benzene. A word of caution: benzene is nasty stuff—poisonous, carcinogenic, flammable—so this experiment should be done in a well-ventilated room. Or use a more benign liquid (e.g., toluene). Note that the sand wetted with benzene is considerably darker than that wetted with water. Yet benzene is just as transparent as water. This is perhaps a puzzle—but it is also a clue. To unravel this puzzle we must first return briefly to single scattering.

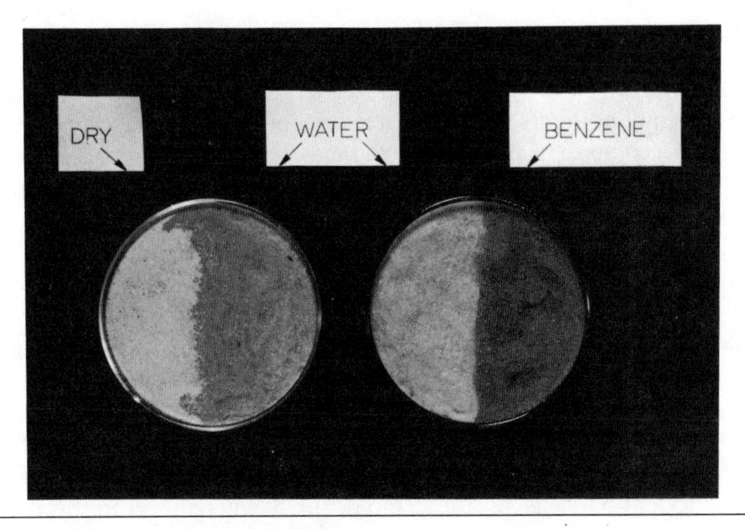

Figure 15.2 Aquarium sand wetted with water and with benzene.

SCATTERING BY A SINGLE GRAIN

Consider an *isolated* scatterer, a sand grain in this instance, although it could be anything. The electric field of a light wave illuminating such a scatterer excites the elementary charges of which it is composed and they consequently radiate, or scatter, a light wave the amplitude of which is not, in general, the same in all directions (for more on this see Chapter 18). It may be helpful to recast this in photon language. Photons incident on a scatterer are scattered in all directions, but more scatter in some directions than in others. We may therefore talk about the probability that an incident photon will be scattered in a particular direction. This, in turn, leads to the concept of the *average* direction in which photons are scattered. This average direction is strongly dependent on, among other things, the size of the scatterer. For example, scatterers much smaller than the wavelength of the light illuminating them scatter nearly the same in all directions. To be more precise, the average scattering angle–the angle between the incident and scattered photons–is 90 degrees: as many photons are scattered into a hemisphere of directions about the forward direction as are scattered into a hemisphere about the backward direction (the forward direction is defined by the direction of the incident beam; the backward direction is opposite this). In contrast, scatterers much larger than the wavelength of the light illuminating them scatter more strongly about the forward direction: the average scattering angle is less than 90 degrees, perhaps 30 degrees or less. Sand grains are much larger than the wavelengths of visible light; although it may not be obvious, this is highly relevant to the puzzle we are trying to unravel.

SCATTERING BY MANY GRAINS

Although we are interested in multiple scattering rather than single scattering, the two cannot be divorced: the characteristics of the individual scatterers (e.g., sand grains) composing an object (e.g., sand) determine its optical properties. To see why we must ask another question: What is it that makes a multiple-scattering medium highly reflecting? Let us assume that it is optically thick; if it is thin, incident photons can leak out the bottom without contributing to the reflected light. In addition to being optically thick, what other characteristics do highly reflecting objects possess? A bright object is one from which most incident photons re-emerge. This in turn implies that not many such photons are absorbed by the object. In each scattering event a photon has a finite probability of being absorbed; stated another way, if a beam of photons is incident on a scatterer, some of them are scattered in all directions and some of them are absorbed. All else being equal an object will be brighter the fewer scattering events incident photons must undergo before re-emerging. And this is where the average scattering angle comes in.

To understand why, consider Figure 15.3 in which I have depicted what happens to photons incident on different objects, one composed of scatterers for which the average scattering angle is 90 degrees and the other for which it is 30 degrees. The two objects are otherwise identical: the probability that a photon is absorbed in each interaction and the average distance a photon travels before being scattered are the same. These paths taken by photons on their way out of the object are representative, other paths are possible.

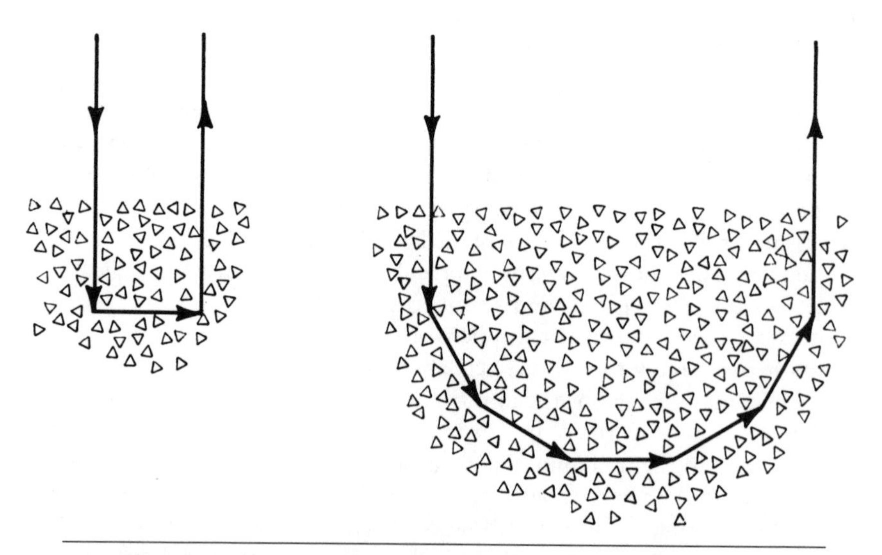

Figure 15.3 The shortest path taken by an incident photon before re-emerging from a collection of many scatterers. On the left the average scattering angle is 90 degrees; on the right it is 30 degrees.

I have chosen the paths shown in Figure 15.3 in such a way that incident photons undergo the *least* number of scattering events before re-emerging. Note that if the average scattering angle is 90 degrees, a photon need be scattered only *twice* before re-emerging. But if the average scattering angle is 30 degrees, a photon must be scattered at least *six* times before re-emerging. The greater the number of scattering events, however, the greater the chance that a photon is absorbed. It is evident, therefore, that when there is absorption the object on the left will be brighter than that on the right if both are illuminated by the same source. Now we have everything at hand to explain Figure 15.2.

I stated previously that the average scattering angle depends on the size of the scatterer; the larger it is compared with the wavelength of the light illuminating it, the more it scatters toward the forward direction. But size is not the only determinant of the average scattering angle; it also depends on what surrounds the scatterers, especially if they are large compared with the wavelength. And it is what surrounds the sand grains—air, water, or benzene—that distinguishes the samples shown in Figure 15.2.

The more closely the optical properties of the surrounding fluid match those of the grains the smaller the average scattering angle. To convince yourself that this is plausible consider the extreme case where the surrounding fluid is optically identical with the grains. In this instance the grains are invisible: the incident light is scattered only in the forward direction. Recall how much more difficult it is to find glassware in water than in air. Water is closer in its optical properties to sand grains than air is. Benzene is even closer yet, which, of course, was why it was chosen for the experiment. Although embedding a single grain in different transparent fluids does not appreciably change the *amount* of incident light it scatters and absorbs, it may appreciably change the *distribution* of the scattered light. A consequence of this change is that the brightness of many grains may be greatly diminished merely by wetting them with different liquids.

COARSE SAND AND FINE SAND

It is implicit in the previous section that there are two characteristics of isolated scatterers which determine what fraction of light incident on an optically thick collection of them is reflected: the average scattering angle and the probability that a photon is absorbed in a single interaction with a scatterer. Beaches provide examples in which one or the other characteristic is the major determinant of what is observed. The striking difference in the appearances of wet and dry sand (Fig. 15.1) was attributed to the first of these characteristics. An example where the second is the determining factor is shown in Figure 15.4, which is a photograph of a dry beach taken by Alistair Fraser. Note that the sand on the right is noticeably darker than that on the left. The water has segregated the sand into two groups, coarse and fine. The coarse sand is darker; this is perhaps more obvious in the closer view of the sand in Figure 15.5.

Figure 15.4 A beach on Kootenay Lake, British Columbia. Note that the sand on the right is darker than that on the left. Photograph by Alistair Fraser.

The average scattering angle for the coarse grains is not appreciably different from that for the fine grains. So it takes about the same number of scattering events for a photon to get out of the coarse sand as to get out of the fine sand. But in each event a photon is more likely to be absorbed by a large grain than by a small grain. As a consequence, the ensemble of large grains is darker.

Figure 15.5 A closer view of the sand shown in Figure 15.4. Photograph by Alistair Fraser.

THE BRIGHTNESS OF FOAM

What is it about foam that makes it so highly reflecting? Foam is a multiple scattering medium; the individual scatterers are bubbles, very thin liquid films surrounding air. The amount of light scattered by a bubble is about the same as that scattered by a water droplet of the same overall size. But because there is so little matter in the bubble it is much less absorbing than the water droplet. A cloud composed of bubbles would, all else being equal, be even brighter than a water droplet cloud. And clouds are usually brighter than beaches. So it is no wonder that foam is highly reflecting: it is made up of scatterers that are very weakly absorbing.

Foam may appear on breaking waves at the beach. And it may also appear on a freshly poured glass of beer that we might bring to the beach. In Chapter 1 I noted in passing that the head on beer is white whereas the liquid from which it is formed is yellow. Why this is so should now be less of a mystery. Light transmitted by the liquid passes through a vastly greater amount of material than does light multiply scattered by the tenuous bubbles in the foam. Beer is yellow because it preferentially absorbs visible light of shorter wavelengths. But there is not enough preferential absorption–indeed, there is very little absorption at all–of the light incident on the foam, so it is white.

16

On a Clear Day You *Can't* See Forever

They can't be more'n ten or fifteen miles. My dad says that's as far as you can see on a clear day.
Dennis the Menace

The quotation at the head of this chapter was the caption for a cartoon that appeared in the local fishwrapper a few years ago. Dennis and a friend are looking at the moon and stars, the "they" that Dennis is referring to. He has drawn an incorrect conclusion from a casual remark made by his father. The moon and stars are much farther than "ten or fifteen miles," and yet on a reasonably clear night they are certainly visible. But during the day, there is an upper limit to the distance over which terrestrial objects are visible, even on the clearest days. It is more than ten or fifteen miles, but not much more. Why the difference between night and day? And why is there an upper limit to visual range? To answer these questions, a simple demonstration will help.

AN AQUARIUM VISIBILITY DEMONSTRATION

In Chapter 14 I discussed a demonstration of multiple scattering in which I used an aquarium. I have used this for yet another demonstration. I filled it with distilled water; on one end I taped a sheet of white paper on which I had pasted three disks, of different size, cut from black paper. I illuminated this white sheet of paper with a slide projector. Then I took my fluorescent desk lamp, the tubes in which are about as long as the aquarium, and illuminated the water from above. Looking through the water along the axis of the aquarium, one sees well-defined black disks against a white background (Fig. 16.1). Then I added a few drops of milk to the water and stirred it thoroughly. The crispness of the disks

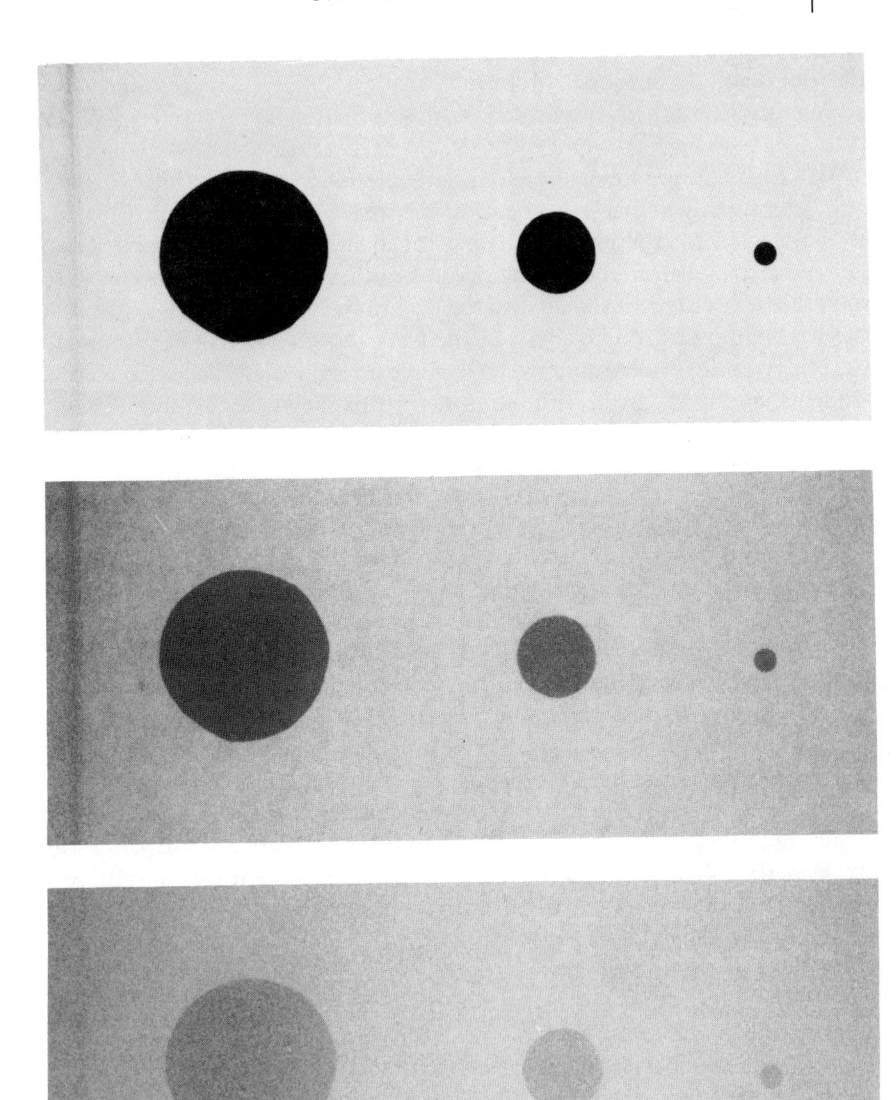

Figure 16.1 Visibility in an aquarium. The three black disks are distinct against a white background when viewed through about a meter of pure water (top). But when a few drops of milk are added to the water the disks become markedly less distinct (middle). With a few more drops, they are even more difficult to distinguish, especially the smallest one (bottom). All these photographs were taken at the same shutter speed and aperture setting.

was noticeably diminished. More milk made them even more difficult to distinguish against their background. With sufficient milk, they were no longer visible.

When you do this demonstration, note carefully how the angular size of a disk determines whether it can be seen; the smallest one disappears first. It is also instructive to experiment with different aquarium bottoms. I painted one side of a rectangular sheet of plastic black, the other white; its dimensions were such that it could be fit snugly into the aquarium. With the black side of the sheet facing upward, the visibility of the black disks is better, not stunningly so, but certainly evident upon careful observation. I tried this out on a few students, and they agreed with me—but perhaps they were just intimidated.

ATMOSPHERIC VISIBILITY

This demonstration is a scaled down version of what happens in the atmosphere. My black disks were less than a meter away. Yet it is not *physical* distance that determines visibility but rather *optical* distance. In any material medium—the atmosphere or my aquarium, for example—there are scatterers: molecules and particles. The scattering mean free path is defined as the average distance a photon travels in the medium before it is scattered. The *optical thickness* of a path connecting any two points in the medium is the distance between them measured in units of mean free paths. Optical thicknesses are dimensionless (i.e., they are measured not in yards or meters or furlongs, but rather are ratios) and may depend on wavelength because the probability of scattering depends on wavelength. A given optical thickness can be obtained with weak scatterers distributed over a long distance (the atmosphere) or strong scatterers distributed over a short distance (aquarium). Physically, the distances are quite different; optically, they are the same.

What is demonstrated, using an aquarium filled with water to which varying amounts of milk are added, can be observed in the atmosphere when you look at a series of parallel ridges, one behind the other. Even though each of them may be covered with more or less the same dark vegetation, their appearance depends on their distance. Each successive ridge in Figure 16.2 is less distinct than its predecessors. Indeed, this is how we judge distances to faraway objects of unknown size. Sometimes we can be deceived when suddenly placed in an environment to which we are not accustomed. This happened to me in Iceland, with nearly tragic consequences.

On a good day (i.e., one on which it is not pelting rain), the air in Iceland can be remarkably clean. During the summer of 1976, when I lived in Wales, my wife and I and our climbing partner, George Greaves, a Scottish mathematician, set out to climb the highest peak in Iceland, the Hvannadalshnukur. It is the highest point on the rim of an extinct volcano, Öraefajökull. When we reached the crater, the Hvannadalshnukur was so distinct that we expected to

Figure 16.2 All these ridges in Central Pennsylvania are covered with the same dark vegetation. Yet they do not appear the same because of airlight: light scattered by all the molecules and particles along the line of sight.

be on top of it shortly. But as we slogged across the flat, snow-filled crater, our goal stayed maddeningly out of reach. Because of the extreme clarity of the air, we had greatly underestimated its distance. In the environment to which we were accustomed, an object so clearly outlined against its background would have been much closer. We no sooner reached the summit when the weather changed suddenly, a common occurrence in Iceland. We unfurled our flag, the Welsh red dragon, took the obligatory summit photograph, and then beat a hasty retreat. Clouds swirled across the crater, and by the time we were back in it, they were so thick that we couldn't see more than a meter or two. The crater was free of crevasses, but once we left it we had to wend our way through them down the steeper slopes that led to our camp. Fortunately, we had placed red flags in the more dangerous spots. We were roped together, George leading. Suddenly he stopped. He shouted to us that he could see down into the valley below. My wife and I joined him, and we all stood looking through the mist at dark objects a few thousand meters below us. Or so we thought. Once again we had been fooled, this time in the opposite direction. Instead of looking into a dark valley several thousand meters away, we were standing on the edge of a crevasse, perhaps thirty meters deep, looking into its dark and murky depths.

I am not the only one to have been fooled by exceptionally clear air. In *Venture to the Interior,* Laurens van der Post has this to say about his initial failure to shoot game on his trek through what is now the mountainous African country of Malawi:

> I realized how wise I had been not to shoot earlier on. The light was clearer than I had imagined. I paced the distance to where I had first shot at the buck. I had thought that it was a hundred to hundred and twenty yards. It was six hundred and fifty paces away. And I tell this story against myself because it shows how pure and clear the air was over the Nyika, and how full, easy, and generous its distances. I am not a bad shot, but it took me five days before I shot my first game.

CONTRAST REDUCTION BY AIRLIGHT

Enough of adventure tales. Let's get back to the more prosaic task of explaining in more detail the demonstration and its relevance to observations.

For us to be able to see any object, the *relative* brightness difference between it and its surroundings—the *contrast*—must be above a threshold value. This value is not an invariable constant like the speed of light or the mass of an electron. Because of differences among people, it may vary from person to person. It depends on the angular size of the object (as evidenced by the demonstration), and even the *absolute* value of the brightness. With this in mind, we should take the often-cited contrast threshold of 2 percent to be merely a rule-of-thumb. Nevertheless, whatever the contrast threshold is, it is finite; we cannot distinguish arbitrarily small brightness differences.

One of my favorite photographs showing contrast reduction is shown in Figure 16.3. It was taken in the Antarctic by David Greegor, a biology pro-

Figure 16.3 Whiteout in Antarctica. These helicopters are on the ground, not flying in clouds. The boundary between ground and sky is indistinguishable. Photograph by David Greegor.

fessor at Nebraska Wesleyan University. Unless you happen to notice that the rotors are tied down, you might think that these helicopters are flying. But they are parked on snow-covered ground. Where is the boundary between ground and sky? The two are indistinguishable because of insufficient contrast. If you have ever skied or driven in a *whiteout* like this, you know how terrifying it can be. Whiteouts are extreme examples of contrast reduction. Less extreme examples are provided by any distant object, such as the ridges shown in Figure 16.2. We can see four ridges, the ones farthest away being the least distinct. There might very well be a fifth one, or a sixth. But even if they were physically present we might not be able to distinguish them because of insufficient contrast. That is, the most distant ridges might be in a whiteout, indistinguishable from the horizon sky. You can sometimes observe this by looking down a single ridge. From the windows of my building I can see a long ridge covered with dark vegetation. Those parts of it closest to me are usually distinct. But often I can look down the ridge to a point where it fades into the horizon sky.

When you look at any object you receive not only light from it but light scattered by all the molecules and particles along your line of sight as well. It is this *airlight* that reduces contrast. Quite often, the brightest object in our field of view is the horizon sky. Even perfectly black objects, at some finite distance, cannot be distinguished against the horizon sky. Let us suppose that it is a clear day. Every molecule and particle along the line of sight between us and a black object is illuminated uniformly by sunlight (see Fig. 16.4 on page 126). Some of this light is scattered toward our eyes, thereby contributing to the airlight we receive. The farther the object, the greater the brightness of this airlight, although it does not increase linearly with increasing optical thickness. Light scattered toward us must not be scattered again if it is to reach our eyes, but the greater the optical thickness the more likely it is to be scattered again. As a consequence, the airlight at first increases linearly with increasing optical thickness; that is, equal increments of optical thickness yield equal increments of brightness. But with ever greater optical distances the additional brightness increment is less and less.

Even in a perfectly clear atmosphere, the optical thickness of a path toward the horizon is effectively infinite (i.e., the brightness of the horizon sky is negligibly different from what it would be if the horizon path were optically infinite), although the brightness of the horizon is finite. The optical thickness of the path between an observer and a black object depends on its distance and the nature and concentration of scatterers along the line of sight. When these combine to give an optical thickness of about 3.9, the relative brightness difference between a black object and the horizon sky is about 2 percent, which we may take to be a good representative value for the contrast threshold. Thus a black object at an optical thickness of 3.9 falls just short of being distinguishable from the horizon sky. Of course, if the object is not black, the optical thickness beyond which it is not visible is even less because light reflected by it adds to the light received by an observer, hence reduces contrast.

In an atmosphere completely free of particles, a sea level physical distance

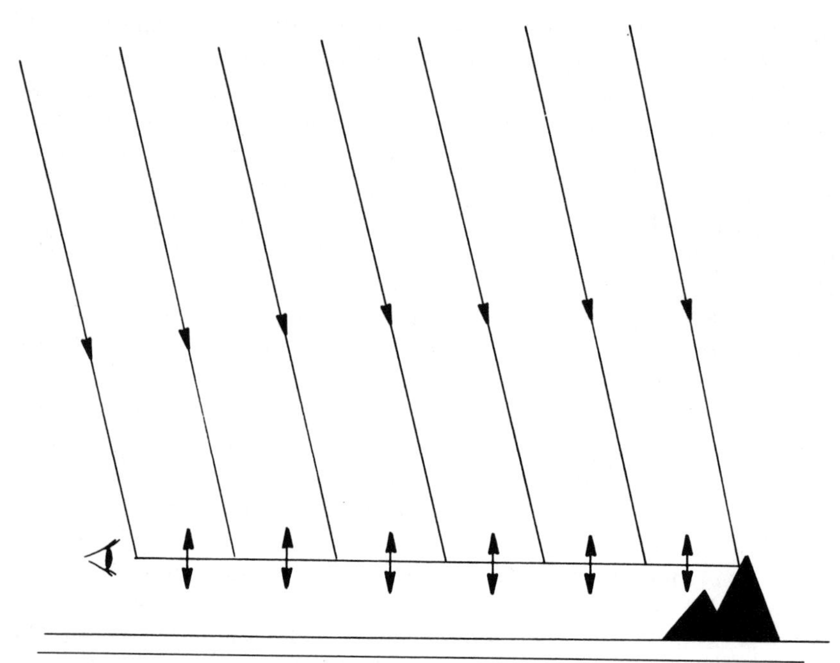

Figure 16.4 An observer who looks at a distant object sees sunlight scattered by everything—molecules and particles—lying along the line of sight. To contribute to airlight, light scattered toward the observer must make it to his eyes without being scattered again. As a consequence, airlight does not increase indefinitely with increasing distance but reaches a limiting value.

of about 330 km on a flat earth corresponds to an optical thickness of 3.9 near the wavelength of peak sensitivity of the human eye. This is the greatest possible *meteorological range,* the distance at which a black object is just visible at sea level assuming a contrast threshold of 2 percent. Of course, meteorological ranges in the real atmosphere (as opposed to an ideal one) will be much less than 330 km because scattering by particles contributes to airlight. According to the International Visibility Code reproduced on page 43 of *Optics of the Atmosphere* by Earl J. McCartney, a meteorological range of greater than 50 km (31 miles) is classified as "exceptionally clear."

A FEW MISCONCEPTIONS TO BE CORRECTED

Attenuation (or extinction; see Chapter 11) *per se* is not what reduces the visibility of distant objects. I sometimes find this explanation given either explicitly or implicitly. In refutation, I note that most stars can't be seen during the day, although they may be seen at night. Attenuation of their light is the same day and night. They can't be seen against the sunlit sky because of insufficient contrast, not because of attenuation. A black object provides no light to be attenuated, and yet its visual range is limited. At some finite distance, airlight will be sufficiently bright that there will no longer be enough contrast between even a black object and the horizon sky. That part of the demonstration in which I changed the bottom of the aquarium also provides evidence to refute the notion that attenuation of light from objects is what diminishes their visibility. When the bottom was changed from white to black, the disks became more visible, even though the background illumination (i.e., the slide projector) remained constant as did the optical thickness. Thus with no change in attenuation by the scatterers along the line of sight, visibility improved. Direct light from the fluorescent lamp as well as light reflected by the white bottom is scattered by the milk particles thereby contributing to contrast reduction. But when the bottom is changed to black, only scattered lamp light contributes to it.

Visibility can deteriorate markedly on very humid days. This is often attributed to water vapor. But even on an extraordinarily hot and humid day, only a few percent of the molecules in the atmosphere are water molecules. Moreover, scattering of visible light by a water molecule is somewhat less than that by either an oxygen or a nitrogen molecule. Water molecules are neither very abundant nor especially strong scatterers compared with the dominant molecular species. Thus water vapor *per se* cannot contribute much to airlight and, hence, to reduction in visual range. It is only when the water vapor *condenses* (i.e., it ceases to be vapor) that it affects visibility. I hope it is clear from Chapter 2 that small, soluble particles in the atmosphere will accrete liquid water when the relative humidity is sufficiently high, but well below 100 percent. When such particles grow, they scatter much more light, hence contribute much more to airlight.

A PARTING SHOT

I could point out some of the subtleties of visibility, but to do so might deprive you of the pleasure of figuring things out for yourselves. From what I have written, you should be able to interpret some observations. For example, one might expect the visual range to be low on cloudy days, and indeed it often is. But in State College, it is on such days that the visual range is greatest. Although on a clear day you can't see forever, you might be able to see farther on a cloudy day than on a clear one. I have observed this so often that I associate exceptionally good visiblity with cloudiness. My guess is that wherever you live, you will observe the same thing. I leave it to you to explain why.

17

A Serendipitous Iridescent Cloud

Let us learn to write essays on . . .
a coloured cloud.
G. K. Chesterton

Several of the demonstrations I have written about have been the result of serendipity. This exotic word comes from a former name for Sri Lanka (also formerly called Ceylon), Serendip. There were three mythical princes of Serendip who often found by chance valuable things they were not seeking. This sometimes happens to me: while doing one demonstration I stumble upon another. For example, it happened once while I was doing the cloud-in-a-bottle demonstration described in Chapter 2.

I usually illuminate the cloud from the side with a photoflood lamp, but since it had been stolen, I had to make do with something less suitable, an overhead projector. Another departure from what I had done in the past was to use a completely clear bottle rather than a partly blackened one. The two sources of light differ in their directionality: the lamp gives light in many directions whereas the projector, as its name implies, gives a beam confined to a narrower set of directions. In one wrestling match with the projector—designed for one purpose, it stoutly resists being used for another—I illuminated the cloud from behind. Immediately there came a cry from the students because they could see colors that I couldn't: without trying I had made an iridescent cloud.

Before explaining the origin of this iridescence, I should like to discuss two demonstrations, the first of which is just a refined version of what I stumbled upon because someone had stolen my lamp.

AN IRIDESCENT CLOUD DEMONSTRATION

This demonstration is a variation on that described in Chapter 2, although its aim is to show not how clouds are formed but rather how they give rise to some optical phenomena in the atmosphere.

Figure 17.1 A cloud formed in a bottle signals its presence by the light it scatters. Under proper conditions, this scattered light exhibits different colors in different directions.

The bottle in which the cloud is made is clear, and the source of illumination is a slide projector into which an opaque slide with a hole in it is inserted; this will give a fairly well-defined beam (Fig. 17.1). To prevent condensation of water droplets on the walls of the bottle, you can add a few drops of liquid detergent to the water in the bottle and then swish it around so that a thin film coats the walls (see Chapter 7). This was suggested to me by one of my students, Tim Nevitt.

As you may recall, a cloud alternately forms and disappears in the bottle when you pressurize and depressurize it. This cloud is evident because of the light scattered by the water droplets. If you are patient, however, you may see more than just a bright pencil of illuminated droplets: you may see some striking colors, especially when looking toward the incident beam, even though the illumination is white light. But don't expect instant success; you'll probably have to make and destroy many clouds. On the first try a dense cloud usually forms and good colors are not seen, especially if there are many condensation nuclei in the bottle. These nuclei are so small that they settle slowly, but when they are incorporated into much larger water droplets they settle comparatively rapidly. With time, therefore, the cloud becomes less dense because nuclei are washed out of the air in the bottle. It is then that the colors are best, although they may last only fleetingly immediately after the cloud is formed (of course, this demonstration should be performed in a darkened room). With the further passage of time the colors deteriorate. Tim and I did a lot of fiddling, and I advise you to do the same. You'll just have to experiment.

If you perform this demonstration before a large group, a tangle of trampled bodies can be avoided by directing the incident beam onto a mirror before it illuminates the bottle. In this way you can control the beam so that everyone can see it without having to move.

Although I can't give you an easy-to-follow recipe for producing striking iridescent clouds in a bottle, I can give you such a recipe for something quite similar but much longer lasting.

SUBCOOLED SULFUR DROPLETS

For this demonstration you'll need two readily available and inexpensive chemicals: sodium thiosulfate (photographic fixer) and hydrochloric acid. Prepare a 0.1 M solution of sodium thiosulfate (24.8 g dissolved in one liter of distilled water) and 0.2 M hydrochloric acid (16 ml of concentrated HCl diluted to one liter). Then add 10 ml of each to 980 ml of distilled water. The resulting dilute solution is initially indistinguishable from distilled water. But with time a subtle transformation occurs, one that you will be unaware of until you illuminate the solution with a beam of white light. After several hours you will see striking colors if you look toward the beam. Moreover, although these colors will change with your vantage point, they will not change rapidly with time. What has happened is that elemental sulfur—one of the components of the solution is sodium *thiosulfate*—has precipitated as small droplets. Although the melting temperature of amorphous sulfur is about 120°C (248°F), the precipitated sulfur is in the form of *liquid* droplets at room temperature, hence they are subcooled, just as many cloud droplets in the atmosphere are subcooled (see Chapter 22).

More details about this experiment are given in a paper by Irving Johnson and Victor LaMer (*Journal of the American Chemical Society*, 1947, Vol. 69, p. 1184), which I recommend for its readability. In particular, they discuss the factors influencing the growth rate of droplets.

Concentration is the most important factor, so to ensure success you should carefully follow the directions in the previous paragraphs. Depending on temperature, the drops grow to sizes around 0.3–0.5 μm in 4–12 hours. So these tiny sulfur droplets are appreciably smaller than cloud droplets, which typically have radii around 10 μm.

Suspensions of small water droplets in air and of small sulfur droplets in water give striking colors near the forward direction when illuminated by a beam of white light. In the following two sections I shall try to explain why; in the final section I shall connect demonstrations and theory with what is observed in the atmosphere.

FRAUNHOFER DIFFRACTION THEORY

Diffraction is one of those technical terms used in so many ways that we might be better off without it. Useful terms are those with unambiguous meanings, ones upon which everyone agrees. Such is not the case with diffraction. It is sometimes used to indicate *any* departure from straight-line propagation of light (i.e., light travels in straight lines except when it doesn't). But it is also sometimes merely a synonym for scattering, especially scattering near the forward direction (i.e., the direction of the incident beam). Or it is used to denote scattering by flat objects, although why such objects diffract light rather than scatter it is a mystery to me. Although I would prefer to toss *diffraction* onto the scrap heap, it is pointless for me to inveigh against a word with such a long history. This need not, however, stop us from recognizing that there is no fundamental difference between diffraction (or reflection; or refraction) and scattering. No one can devise an experiment to distinguish between a diffracted photon and a scattered photon.

The colors exhibited near the forward direction when small droplets are illuminated by white light could be—indeed have been—denoted as a diffraction phenomenon. What is observed would in no way be changed if scattering were substituted for diffraction. Nevertheless, there may be good reasons for using the term diffraction provided we do so intelligently. After years of meditation, I have arrived at the following recommendation: we agree to call something a diffraction phenomenon if to our satisfaction it can be described by an *approximate* theory called diffraction theory (better yet, Fraunhofer diffraction theory). Beyond this we will not go. We will not dissipate our energy on senseless arguments about whether particles really diffract or scatter. The distinction here (one that is sometimes not made) is between an observable phenomenon and the hierarchy of theories by means of which we try to make some sense out of it.

Fraunhofer diffraction theory does not *cause* iridescent clouds, it merely provides an *approximate* theoretical framework for understanding them. There are other such frameworks, an indefinite number of them. In general, they are more precise at the expense of increased complexity. If we are smart—life is short, after all—we choose the simplest one for our purposes, in this instance Fraunhofer diffraction theory.

I cannot discuss this theory in detail, but I can at least try to convey its flavor. According to Fraunhofer diffraction theory, scattering by any particle is identical with scattering by its projection onto the incident beam (e.g., scattering by a sphere is identical with scattering by an opaque circular disk). Moreover, this theory is oblivious to the composition of the particle and the polarization state of the incident light (see Chapter 19), clear indications that the theory can be only approximate.

Now suppose that a wave is incident on an opaque disk. This disk partially blocks the incident wave. Each part of the unblocked wave in the plane of the disk is looked upon as the source of secondary waves all of which combine to give the wave beyond the disk (i.e., the wave scattered by it).

Fraunhofer diffraction theory is quite simple in principle; it entails no more than adding waves together. Because it is simple it has its limitations. In particular, it is limited to scatterers much larger than the wavelength of the illuminating light and to scattering near the forward direction. Fraunhofer diffraction theory fails completely to account for the blueness of the sky (caused by scattering by molecules, which are much smaller than the wavelengths of visible light). Nor does it reproduce any of the features of rainbows, which occur well away from the forward direction.

SCATTERING BY A CLOUD DROPLET

Any particle illuminated by a beam of light scatters light in all directions, but in some directions more than in others. The angular dependence of scattering by a typical cloud droplet (radius = 10 μm) near the forward direction, calculated according to Fraunhofer diffraction theory, is shown in Figure 17.2. I could have used a more complicated theory to obtain this figure, but it would have been like shooting a mouse with an elephant gun.

Note that scattering is highly peaked near the forward direction, which is characteristic of particles larger than the wavelength of the light illuminating them; the larger the particle, the more strongly it scatters near the forward direction (see the following chapter). Note also that away from this direction there are narrow angular ranges over which scattering is much greater than over others. Moreover, the positions of these peaks in the scattering diagram depend on the wavelength of the incident light. For example, if the incident light is green, the

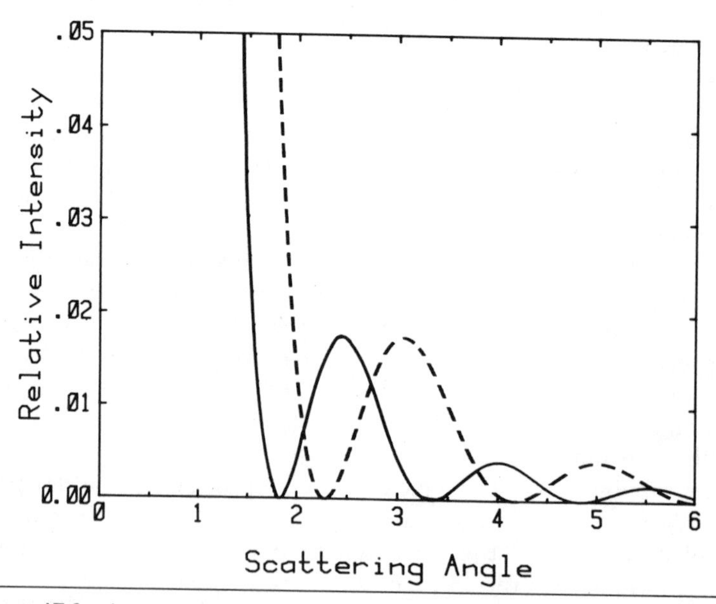

Figure 17.2 A small water droplet (radius 10 μm) scatters light in all directions but in some more than in others. The scattering diagram for red incident light (broken line) is shifted relative to that for green (solid line), hence providing a mechanism for color separation of white light.

first peak in the scattering diagram occurs at about 2.5 degrees; but for red light, the peak is shifted about a degree. Regardless of the wavelength, the *total* amount of light scattered in all directions is about the same, but its distribution is different, which is sometimes observable.

Suppose, for example, that a 10 μm cloud droplet is illuminated by a beam of white light, a mixture of all colors. The angular distribution of each component scattered is different. For example, the intensity of scattered red light at about 3.5 degrees is much greater than that of green light (see Fig. 17.2). If we were to look in this direction we would see red light. In another direction, 2.5 degrees say, the scattered light would be greenish. What we see depends on where (i.e., in which direction) we look. The particle is a kind of angular filter: in some directions, it scatters light of some colors more than others thereby separating the colors of incident white light. The angles at which the scattered light of a given color is most intense depend on droplet size as does the separation between the intensity peaks for different colors; the larger the droplet the smaller this separation.

Up to this point I have had in mind scattering by a single droplet. But we almost always observe a cloud of them, as in the demonstrations described previously. To see vivid colors in the light scattered at different angles by such clouds requires that several conditions be met: multiple scattering (see Chapters 14 and 15) must be negligible; the incident light must be narrowly directed; all the droplets in the cloud must be about the same size and not too large. Raindrops illuminated by sunlight are much too large to give the kinds of colors I've been talking about. But what about rainbows? Some people—the same ones who call every small bird a sparrow—call every splash of colors a rainbow. But proper rainbows occur well *away* from the sun. What do we see when we look toward it?

CORONAS AND IRIDESCENT CLOUDS

With a thin veil of small water droplets directly between you and the sun you might be lucky enough to see a magnificent *corona:* a series of colored rings around the sun and a few degrees from it. A few qualifications are in order before I proceed. By the solar corona, solar physicists mean something quite different from what I am discussing here. And although I say "water droplets," the particles responsible for coronas could just as well be spherical ice particles or randomly oriented ice needles (in the demonstrations, the droplets were definitely droplets, either water or sulfur). Finally, the source of light could just as well be the moon.

Well-defined colored rings a few degrees from the sun result from scattering by a tenuous collection of droplets all of which are nearly the same size. And just as the rainbow is a mosaic, so also is the corona: each part of it is contributed by different droplets (see Chapter 21 for a further discussion of this). If the droplets differ appreciably in size, there will be a bright *aureole* around the sun, but it will show little color.

The sun is not strictly a unidirectional beam; it is a mixture of beams, directed slightly differently, each of which is scattered by a given droplet. All of these individual scattering diagrams combine to give what is observed. Thus if the droplet size is such that the angular separation between scattering peaks for different colors (red and green, say) is appreciably less than the angular width of the sun (about half a degree), then the result is more or less white light in all directions even if all the droplets are the same size. So if you see a corona you can be reasonably sure that the droplets responsible for it have radii not much greater than about 30 μm. Indeed, if you can measure the angular separation between different rings of the same color, you can estimate the droplet size; this is a rather simple example of what is called remote sensing of the atmosphere, an activity very much in vogue these days.

Lest I mislead you, coronas are not seen often. The few occasions on which I have seen good ones are memorable. This should come as no surprise because the requirements for a perfect corona—a thin cloud of small droplets all about the same size and uniformly covering the sky within few degrees of the sun—are rather strict and hence not likely to be met frequently. But I have seen many, many coronas, not stunning perhaps, but still a source of small delight. These coronas have not been in the sky; they have been on windows and in my breath. For example, one cold winter day I looked through my front door window to see a corona; the light source was a lamp on my lawn and the particles were condensed droplets. Also during the winter I have seen lovely coronas while looking through the fogged windshield of my truck at distant lights; it seemed a pity to turn on the defroster. After steamy baths in seamy motels I have seen coronas while looking through the fogged windows. And I have often amused myself while walking home from work on wintry nights by breathing out mixing clouds (see Chapter 5) and looking at the fleeting coronas produced by the lights of oncoming cars or by distant street lights. Once you know what to look for, you begin to see coronas everywhere. You don't even have to wait for winter. Duncan Blanchard once wrote to me that often he has amused himself by breathing on the window of his front door. Like me, he has a light on his lawn, around which he can see coronas. By breathing more or less, the droplet size changes giving rise to differences in colors. And then he watches the corona change as the droplets on the pane shrink and eventually disappear.

Although coronas are infrequent in the sky, *iridescence* is not. Unlike coronas, the pastel colors of iridescent clouds form no regular pattern; they are merely bits and pieces of coronas, incomplete coronas, perhaps ones caused by droplets of different size (the *Glossary of Meteorology* conveys the notion that iridescent clouds are ice-crystal clouds, but this is by no means necessarily so). The sky around the sun does not have to be uniformly covered with thin clouds. You can see iridescence even at the edges of thick clouds. To avoid being dazzled by sunlight I sometimes scan the sky near the sun by looking at its much less intense reflection in a piece of black glass. Puddles on asphalt serve the same purpose, and following rains I have seen the tinted edges of clouds while looking downward rather than toward the heavens. While driving, especially in winter when the sun is low, I often see scattered iridescent clouds, green and red intermixed with white like neapolitan ice cream.

The most memorable iridescent cloud I have ever seen was truly serendipitous. On a humid summer day I was walking with a friend, Brian Thomason, along an isolated sandy shelf just below a cliff overlooking the sea at Point Loma in Southern California. Not far to the east, there is a naval airfield. Suddenly the silence was shattered by the roar of a huge airplane climbing rapidly as it headed out over the Pacific. Reflexively, perhaps angrily, we looked up at the intruder on our solitude. We gasped in unison at what we saw: vivid iridescence in the cloud formed at the edge of the wing for the brief moment during which the plane passed between us and the sun. Because Brian's spontaneous reaction was similar to mine I knew that what I had seen was real. Frantically, I reached for my camera. But the magic moment had passed. It is permanently imprinted on my memory but not, alas, on film.

18

Physics on a Manure Heap: More about Black Clouds

We are all in the gutter, but some of us are looking at the stars.
Oscar Wilde

Late one spring, an event I had long been awaiting finally occurred. It had rained heavily during the night, but skies were clear by morning. Driven by the rising sun, mixing clouds (Chapter 5) were curling off roofs and trees and lawns and wooden fences. Out I went excitedly, armed with camera, and madly snapped it all. Then I hurried to the fields where I run my dog every morning. Mixing clouds were roiling off of trees and piles of dead leaves. These, too, I snapped, and then moved on, looking for new treasures. Suddenly I saw one: a great heap of steaming manure. It was marvelous; I could hardly control my excitement. I raced around it wildly, looking now in one direction, now in another. Toward the sun, which was still fairly low in the sky, mixing clouds were clearly evident, but as I moved they became fainter. They all but disappeared in the direction away from the sun. First I photographed the manure heap looking toward the sun, then I ran around to the other side and photographed it looking in the opposite direction (Fig. 18.1). It is considered bad form for photographers to include their shadows in their shots, but I deliberately included mine to leave no doubt as to the position of the sun. Note the great difference between two photographs of the same object taken 180 degrees apart. If you think that this has something to do with different contrast between clouds and their surroundings, the photographs in Figure 18.2 of the same cloud-girded tree taken from two opposite directions ought to dispel that notion (see page 138).

What I observed that delightfully steamy morning were just a few examples

Figure 18.1 These mixing clouds on a manure heap were photographed at the same time but in different directions. The top photograph was taken looking toward the sun; the bottom one was taken looking away from it.

of the consequences of the extreme asymmetry between forward and backward scattering by cloud droplets, or indeed by any particles comparable with or larger than the wavelength of the light illuminating them. At the edge of the same field where I found the steaming heap of manure, I encountered yet another example one morning. It had been foggy, and winter was far enough away that spiders were still at work. Evidence of their handiwork festooned a stand of tall and brittle weeds. The webs glistened from the droplets strung from them like beads on a necklace. They glistened, that is, when I looked toward the sun.

Figure 18.2 Mixing clouds on a tree. The left photograph was taken looking into the sun; the right one was taken looking away from it. The background for the mixing clouds roiling from the tree was essentially the same in both photographs, yet in one clouds are evident whereas in the other they have all but disappeared.

When I looked at the weeds from the opposite direction, the webs vanished, and yet nothing had changed except my vantage point.

It is easy enough to give laboratory demonstrations of forward-backward asymmetry in scattering by particles. Just about any particles that you can make readily are larger than the wavelengths of visible light. Examples are water droplets formed by inexpensive atomizers, the kinds that dispense perfume. Talcum powder is another source of particles. And you can make bubbles by rapidly pouring water from one large container into another; they last long enough to observe scattering by them in various directions. Besides particles, all that you need is a collimated source of light, my favorite being a slide projector in which a black slide with a hole in it is inserted. First look at the light scattered near the forward direction by the particles (i.e., look toward the source of light), then look in the opposite direction. The difference is striking. Neither the particles nor the illumination has changed, but what is observed is vastly different. It must be that scattering is not the same in all directions, but rather is much greater toward the forward direction than toward the backward direction. Why?

At this point I could just say rather stuffily that intricate computations, which are well beyond the scope of this book, show unambiguously that the larger a particle is the more it scatters near the forward direction relative to the backward direction. This I would find very unsatisfying, so I shall attempt to give a simple explanation. If you are not interested in a detailed explanation of forward-backward scattering asymmetry, you can skip the next section. I have made my arguments as simple as possible (much simpler than those required for a complete explanation), but to understand them may still require a bit of head scratching.

SCATTERING BY DIPOLES

A dipole, as its name implies, is essentially *two* charges, one positive the other negative, of equal magnitude, and separated by a finite distance. Dipoles are electrically neutral, although they still contain charges. Suppose that a dipole is illuminated by a beam of light, that is, by an electromagnetic wave. The two charges making up the dipole are set into oscillatory motion at the same frequency as that of the incident wave. As I explained in Chapter 10, charges in oscillatory motion—indeed, in any accelerated motion—radiate electromagnetic waves because of the finiteness of the speed of light. Thus when excited by an electromagnetic wave, a dipole radiates, or scatters, such a wave in all directions. Now let us complicate matters slightly by considering two identical dipoles illuminated by the same wave (Fig. 18.3). What is the total wave scattered by the two?

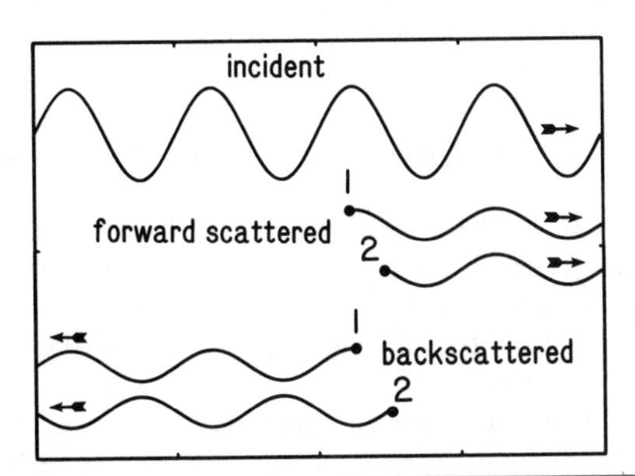

Figure 18.3 Excited by an incident wave, two dipoles scatter waves in all directions. When these two waves are added together, the resultant wave depends not only on the separation of the dipoles, but on the direction of scattering as well. In the forward direction, the two waves are exactly in phase, regardless of the separation of the dipoles. This is not true for any other direction. For the example chosen here (two dipoles one-quarter wavelength apart), the scattered waves are exactly out of phase in the backward direction. Figure courtesy of Roger Johnston.

To answer this question we must add the two scattered waves taking due account of their *phase difference:* the two dipoles are *coherent* scatterers (see Chapters 13 and 15). At some point in space the two waves may be such that at any instant the field of one is equal in magnitude but opposite in direction to that of the other. When we add the two waves, their fields cancel: they are exactly out of phase. They may, however, be exactly in phase, that is, the fields of the two waves are equal both in magnitude and direction, in which instance the total field is twice that associated with a single dipole scatterer.

What most detectors, including our eyes, respond to is not the field but rather the *square* of the field, often called the intensity of the light. When two identical waves are in phase, the intensity corresponding to their sum is *four* times that of the intensity corresponding to a single wave. Two waves out of phase cancel, and therefore give rise to a wave with *zero* intensity. Because of the wave nature of light, the intensity of the sum of two identical waves may be anywhere between four times that of one and zero depending on their phase difference.

Each of the two dipoles shown in Figure 18.3 is excited by the incident wave and each scatters in all directions. To determine the wave scattered by the two we must add the separate waves taking account of their phase difference, which depends on the separation of the dipoles *and* the direction of scattering. To show this dependence I have considered just two directions, forward and backward. The dipoles are separated by one-quarter wavelength. Consider first the forward direction. The two dipoles are excited out of phase because they are separated. But for the same reason, the waves radiated by the two dipoles are also out of phase, and by the same amount. Thus the net effect is that the two forward-scattered waves are in phase. Stated another way, dipole 1 is excited *ahead* of dipole 2; the wave from 1, however, is *behind* that from 2. But note what happens in the backward direction. The two backscattered waves, which are mirror images of the forward-scattered waves, are now exactly out of phase; dipole 1 is excited ahead of dipole 2, and the wave scattered by 1 is also ahead of that scattered by 2.

I have picked a particular separation of the dipoles merely for illustration. Regardless of their separation, scattering by them is always in phase in the forward direction, the only direction for which this holds. In the forward direction, therefore, the intensity of the light scattered by two dipoles is four times that scattered by one; that scattered by four dipoles is 16 times that scattered by one, and so on. In the backward direction, however, scattering does not increase nearly so rapidly with an increase in the number of dipoles because in this direction the dipoles do not necessarily scatter exactly in phase.

To make my point, I have ignored interactions among the dipoles. By interactions, I mean that a given dipole is excited not only by the incident wave, but also by the waves scattered by its fellow dipoles. This interaction complicates matters, but it does not change my result substantially. The essential point is that scattering in (or near) the forward direction is special.

This point is shown in Figure 18.4. The scattered intensity is that for a single dipole, two dipoles, and four dipoles all on the same line and separated by one

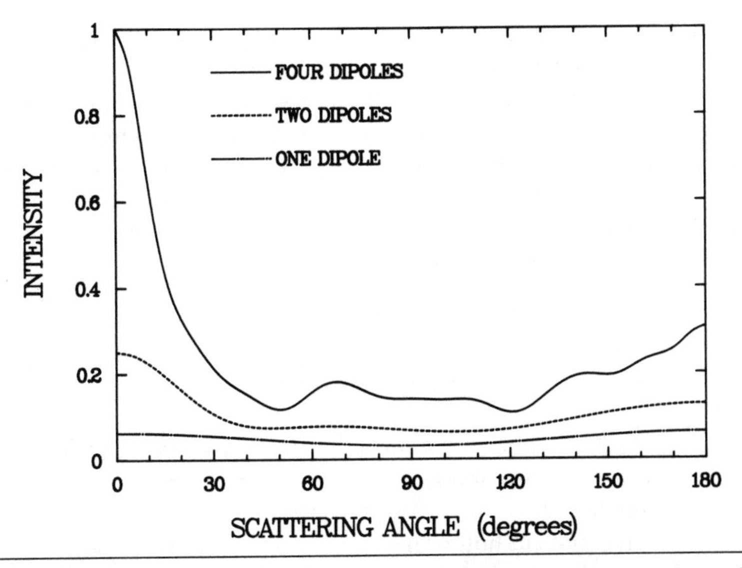

Figure 18.4 The greater the number of dipoles in an array, the more they collectively scatter toward the forward direction. This is evident with only a few dipoles. For the example shown here, all of the dipoles lie on the same line, are separated by one wavelength, and interact with one another. The scattered intensity has been averaged over all orientations of the line of dipoles. Figure courtesy of Shermila Brito Singham.

wavelength. For these exact calculations, the dipoles were allowed to interact, and scattering was averaged over all orientations of the line of dipoles relative to the direction of the incident light. Note that the greater the number of dipoles, the greater the forward-backward asymmetry. And this is evident with only a few of them.

I have chosen dipoles for illustration because the dipole is the simplest (uncharged) scatterer. More important, however, is that any particle becomes an array of dipoles when illuminated by an electromagnetic wave. The molecules making up the particle are uncharged: the amounts of positive and negative charge are equal. When isolated, the center of the positive charge in a molecule may coincide with that of negative charge. In an electric field, however, these two charge centers are displaced slightly because they are forced in opposite directions by the field; a dipole is therefore created or *induced* by the field. If this field is oscillatory, the dipole oscillates and consequently radiates. Thus to determine how a particle scatters light we could, in principle at least, merely add up the waves scattered by all of its constituent molecules (i.e., dipoles) taking account of their phase differences. Indeed, if it were not for interactions among the dipoles, this would be easy, no more than tedious but straightforward bookkeeping. The arguments I made for a few dipoles can be extended to a huge array of them, which is just what any illuminated particle is. Thus the larger the particle (i.e., the more molecules it contains), the more it scatters in the forward direction relative to the backward direction.

If a particle is much smaller than the wavelength of the light illuminating it, the distance between each pair of its constituent molecules must be similarly small. It should be evident from Figure 18.3 that two dipoles separated by distances that are small compared with the wavelength scatter in phase in *all* directions. Thus we expect very small particles to scatter more or less the same in all directions, and this is indeed what is observed.

BLACK CLOUDS

The subtitle of this chapter is "More about Black Clouds." In Chapter 11, I discussed one reason why clouds are black. Now that I have laid the foundations, I can discuss yet another reason.

One day some students hunted me down in my office. They had been watching from our weather station scattered clouds on an otherwise clear day in late fall; the sun was not high in the sky. What they saw puzzled them. All the clouds were more or less the same, and yet some were brighter than others. Those seen looking toward the sun were brighter than those seen looking away from it. I went up onto the roof and took the two cloud photographs shown in Figure 18.5, first looking toward the sun and then directly away from it. These clouds are merely a large-scale version of the mixing clouds on the manure heap with which I began this article. The asymmetry in their brightness is a consequence of the extreme forward-backward asymmetry in scattering by cloud droplets. I must emphasize that this asymmetry manifests itself only in thin clouds. Don't look for it in great towering cumulus clouds. When you examine carefully the clouds away from the sun in Figure 18.5, you will notice that although they are certainly darker than those toward it, they all are not equally dark. The biggest of these dark clouds are also the comparatively brightest ones. Why this is so I leave for you to figure out for yourselves (you'll find some hints in Chapter 14).

A sequel to the story in the preceding paragraph is that I put the question that the students asked me on a subsequent examination. Some of them got it.

Once you have become aware of the extreme forward-backward asymmetry in scattering by particles, you will stumble across many examples of it. For example, in my yearly journeys across the United States, somewhere in Nebraska I begin to encounter fields irrigated by great gangs of sprinklers spitting clouds of droplets. As I drive by, the brightness of these clouds often changes dramatically, especially if the sun is low in the sky. The illumination is constant, the droplets are constant, even the contrast with the surrounding fields is constant. All that changes is my line of sight, yet this leads to readily observable consequences.

Figure 18.5 All the clouds shown in these photographs are more or less the same. Yet those seen looking toward the sun (top) are brighter than those seen away from it (bottom). This is a consequence of the extreme asymmetry in forward-backward scattering by cloud droplets.

19

Polarization of Skylight

Have not the Rays of Light several sides endued with several original Properties?
Isaac Newton

Everyone, except perhaps residents of the most polluted cities, knows that skylight is blue. Yet it has been my experience that few people know that it is partially polarized. This is hardly surprising because blue skylight can be observed effortlessly. Indeed, it would take a conscious act of will *not* to observe it when present. But to observe polarization of skylight requires more than just glancing skyward. One must know what to look for and where. Moreover, it is difficult to observe polarization with only the unaided human eye: one needs polarizing filters. Such filters are incorporated in many sunglasses, and with their aid one can scan the sky for patterns that are hidden to the unaided eye.

A POLARIZATION DEMONSTRATION

Before grappling with the abstract concept of polarized light, a simple demonstration may provide some incentive for doing so. Fill a jar with clean water and add a few drops of milk to it. Illuminate this milky water in a dark room with a narrow beam of light, which can be obtained with a slide projector and an opaque slide with a hole in it. Fat globules in the milk scatter light in all directions resulting in a column of light like that shown in Figure 19.1. Look at the light scattered perpendicular to the incident beam through a polarizing filter such as polarizing sunglasses. While you rotate this filter the intensity will wax and wane, probably even vanish altogether. Yet if you look directly into the beam the intensity will not change as you rotate the filter. Because of its interaction with the fat globules, the incident light has been transformed: although derived from it, the scattered light has somehow acquired different properties that we would not have noticed without the aid of the polarizing filter.

I suspended the slide projector so that the beam was vertical merely for

Figure 19.1 The column of light is light scattered by a dilute suspension of milk in water. When viewed perpendicularly through a polarizing filter this column can be made to disappear by rotating the filter.

photographic purposes. For demonstrations before audiences a horizontal beam is more convenient. Although it is usually not possible to provide everyone with polarizing filters, it is not necessary to do so: merely rotate a filter in the beam *before* it illuminates the water. As the filter is rotated the scattered light will wax and wane. But if the demonstrator's line of sight is vertical (i.e., looking into the mouth of the jar) whereas the audience's is horizontal, the sequences they observe will not be the same: when the scattered light is most intense for the demonstrator it is least intense for the audience and vice versa. Moreover, the extremes are obtained each time the filter is rotated 90 degrees. These are clues about polarized light and its transformations. Now we are ready to probe deeper into the nature of light. Indeed, we have little choice but to do so if we are to understand the demonstration.

THE NATURE OF POLARIZED LIGHT

Light is an electromagnetic wave, oscillating electric and magnetic fields traveling in concert (see Chapter 10), the discovery of which more than a century ago was one of the greatest triumphs of the Scottish physicist James Clerk Maxwell. In air the electric and magnetic fields of a light wave oscillate perpendicular to its direction of propagation, just as the expanding ripples produced by raindrops falling on a pond undulate perpendicular to its surface. It has become customary in recent years to specify the state of polarization of light by the behavior of the electric field; this is sufficient because the magnetic field is perpendicular to the electric field.

Subject only to the restriction that it be perpendicular to the direction of propagation, the electric field may oscillate in many distinguishable ways. It may always point in a single direction, for example, in which instance the light is said to be linearly polarized. There are many more possibilities, but for our purposes it will be sufficient to distinguish between unpolarized and linearly polarized (both completely and partially) light.

We may consider any wave to be a superposition of two waves linearly polarized at right angles to each other. The electric field strength of these two waves may fluctuate. If one field fluctuates independently of the other, and if the (average) intensities of the two waves are equal, the light is said to be unpolarized. A consequence of this is that the various directions of the total electric field as it rapidly oscillates form no well-defined pattern. If such a pattern exists the light is completely polarized. An example is linearly polarized light: the electric field always points in a definite direction perpendicular to the direction of propagation. By way of contrast, the electric field of an unpolarized wave has no preferred direction: now it points in one direction, now in another. Sunlight and light from incandescent sources such as slide projectors are unpolarized.

Partially polarized light lies anywhere between completely polarized and unpolarized light. We may consider any beam of light to be a superposition of two beams, one unpolarized, the other completely polarized. The ratio of the intensity of the polarized component to the total intensity is called the degree of polarization. Completely polarized light is often said to be 100 percent polarized.

Two beams, identical in all respects except polarization state, can nevertheless interact differently with matter. If it were not for this we would either be ignorant of polarization or indifferent to it.

POLARIZATION UPON SCATTERING

I showed in the previous chapter that the intensity of light scattered by a particle is generally not the same in all directions. The extent to which it varies depends on, among other things, the state of polarization of the incident light. An analogy may help to explain why.

The apparent length of a pencil may change as we look at it from different directions all of which lie in the same plane. If the pencil is perpendicular to this plane its length does not change. But if it lies in this plane its length varies with the direction of observation. Indeed, the pencil reduces to a point when we look along its axis.

Consider now what happens when linearly polarized light illuminates a particle much smaller than the wavelength. Because of the smallness of the particle all of its charges oscillate in phase: they are excited at the same time by the same field. And if the particle is spherical we expect them to oscillate along a direction parallel to the incident electric field. As far as scattering of light is concerned, the particle is a pencil of oscillating charges, called a dipole, which radiates an electromagnetic wave in all directions (for more on dipole radiation see the previous chapter).

Suppose that the incident light exciting a dipole is linearly polarized, either horizontally or vertically. Horizontal and vertical have no meanings until a reference plane is chosen. Two lines determine such a plane, so a logical choice is that determined by lines in the directions of the incident and scattered waves, the scattering plane. As you scan in this plane the scattering angle—the angle between the incident and scattered waves—varies between zero degrees in the forward direction, the direction of propagation of the incident wave, and 180 degrees in the backward direction, the direction opposite that of propagation.

When you scan in a horizontal plane the intensity of the light scattered by a small sphere does not vary with scattering angle if the incident light is vertically polarized. But if the incident light is horizontally polarized, the scattered light is most intense in the forward and backward directions and vanishes at 90 degrees to the incident beam; that is, the scattered light varies in much the same way—for much the same reason—that the length of a horizontal pencil varies when viewed from different directions lying in a horizontal plane.

This demonstration is consistent with the demonstration in which a polarizing filter is rotated in the beam before it illuminates a dilute milk suspension. The filter linearly polarizes the unpolarized light from the projector. Thus it is polarized light that is scattered by the fat globules, which are smaller than the wavelengths of visible light. When viewed at 90 degrees to the incident beam, the scattered light can be made to vanish if the incident light is horizontally polarized. Recall that the demonstrator and the audience do not see the same sequence of events. It should now be evident why: if he looks into the mouth of the jar while the audience looks at its side, what is horizontally polarized for him is vertically polarized for them. When the scattered light vanishes for him it is most intense for the audience and vice versa. Moreover, the intensity maxima and minima correspond to filter positions rotated through 90 degrees.

Now all the ingredients are at hand to explain why unpolarized light may be transformed into partially polarized light upon scattering. An unpolarized beam is a mixture of two independent, linearly polarized beams of the same intensity, horizontally and vertically polarized, say. Except in the forward and backward directions, these two beams are scattered differently by a small sphere. This difference is greatest at a scattering angle of 90 degrees, where the intensity of the scattered light vanishes for the horizontally polarized beam. At this angle, therefore, the scattered light is completely vertically polarized. At other scattering angles there is some mixture of horizontally and vertically polarized light, although the horizontal component is less intense. Hence the polarization is less than complete: the scattered light is partially polarized perpendicular to the scattering plane. The degree of polarization of initially unpolarized light scattered by a small sphere is shown as a function of the scattering angle in Figure 19.2.

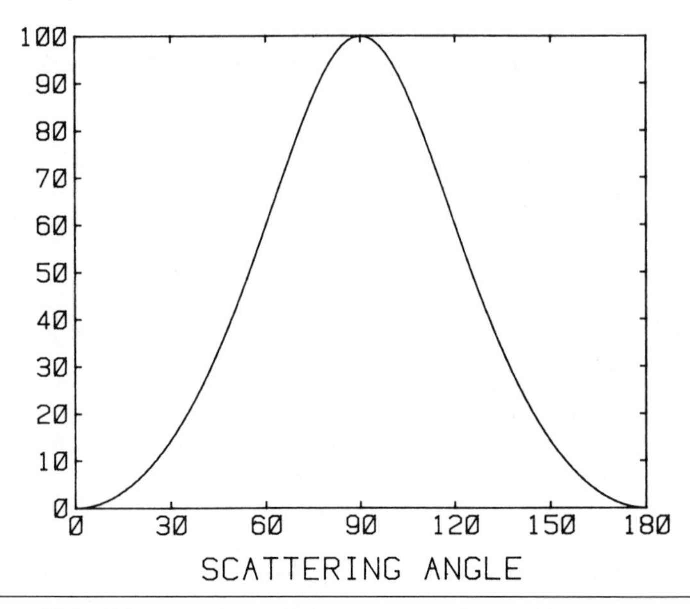

Figure 19.2 When unpolarized light is scattered by a small, spherical particle the result is partially polarized light. Depending on the scattering angle—the angle between incident and scattered beams—the degree of polarization varies symmetrically about 90 degrees from 0 to 100 percent.

Clear air contains many small scatterers, the air molecules themselves. As a consequence, skylight is partially polarized because unpolarized sunlight acquires a degree of polarization upon scattering by air molecules. This can be observed by scanning the sky through a polarizing filter while rotating it. The degree of polarization will be greatest at 90 degrees to the sun, although it will never be 100 percent for reasons I shall discuss later.

Several conditions underlie the curve in Figure 19.2. One of them is that the scatterer must be small compared with the wavelength of the light illuminating it. In general, unpolarized light scattered by larger particles will not be highly polarized at 90 degrees. This can be demonstrated by mixing some large particles–I have used chalk dust–into the dilute milk suspension. Now when the polarizing filter is rotated, the column of light does not completely disappear, as shown in Figure 19.3. Light scattered by the small fat globules can be made to vanish, but not light scattered by the much larger chalk particles.

The difference between the polarizing properties of small and large scatterers can be put to good use in photography, particularly of clouds. Cloud droplets are much larger than the wavelengths of visible light; and they are not all the same size, especially in cumulus clouds. As a consequence, they do not highly polarize the light they scatter. Hence contrast between sky and clouds can be enhanced by using a polarizing filter, as I did when I took the photographs shown in Figure 19.4 of clouds over the Gila Wilderness in New Mexico.

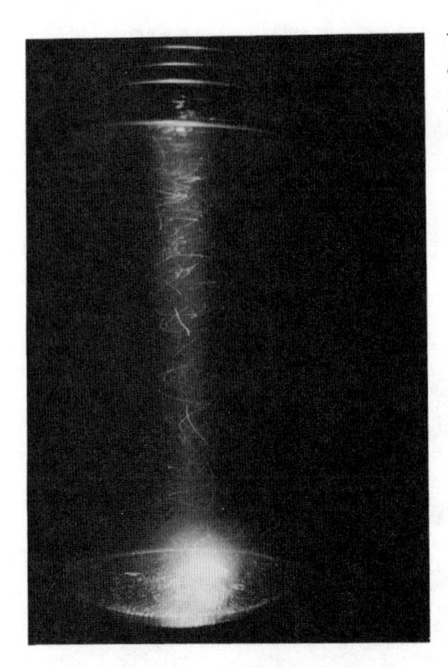

Figure 19.3 When chalk dust is added to the jar of milky water shown in Figure 19.1 the column of light cannot be made to disappear completely, no matter how the polarizing filter is rotated. The streaks are caused by particles moving during the time of exposure.

One photograph shows what the scene looked like to the unaided eye. When I took the other photograph I rotated the polarizing filter on my lens until contrast was greatest. A polarizing filter can reduce the intensity of any light source by at least half (you can verify this by noting exposure times with and without a filter). If the source is partially (linearly) polarized its intensity can be reduced by more than half depending on the orientation of the filter. Both skylight and cloudlight seen through a polarizing filter are reduced in intensity, but partially polarized skylight is reduced more than unpolarized cloudlight, which enhances contrast between clouds and sky.

There are subtleties to be seen when viewing clouds through a polarizing filter. Not only is contrast between clouds and sky enhanced, clouds, especially those most distant, take on a yellowish or reddish hue. Look at near and distant clouds through polarizing sunglasses while rotating them. Note the subtle color changes.

I pointed out in Chapter 16 that when you look at any distant object, such as a cloud, you receive not only light from that object but airlight as well. Without airlight a cloud would appear redder the farther away it is, even with the sun directly overhead. A distant cloud is much like the sun on the horizon: the light you receive is light that has not been scattered, hence it is reddened because the shorter wavelength components of white light have been preferentially scattered *out* of your line of sight (see Chapter 12). Airlight, however, is light that has been scattered *toward* your eye, hence it is bluer than transmitted cloudlight. The two combine so that the cloud does not appear as red as it would if there were no airlight. But airlight, unlike cloudlight, is partially polarized so it can be partly eliminated by using a polarizing filter. This will tip the balance somewhat in favor of the redder cloudlight.

Figure 19.4 The top photograph of clouds over the Gila Wilderness in New Mexico shows the contrast between sky and clouds as it appears to the unaided eye. The bottom photograph was taken through a polarizing filter rotated so as to obtain the greatest contrast.

DEPARTURES FROM COMPLETE POLARIZATION

In addition to being strictly applicable only to particles much smaller than the wavelength of the light illuminating them, the polarization curve in Figure 19.2 is further restricted to spherical scatterers. With the exception of argon, the major molecular constituents of the atmosphere are not spherically symmetric. This reduces the maximum degree of polarization of skylight from 100 percent to about 94 percent. But this upper limit is unlikely to be observed, especially near sea level.

Even if the scatterers are small and spherical, multiple scattering (see Chapters 14 and 15) will further reduce the maximum degree of polarization. You can demonstrate this with a milk suspension. If it is very dilute then at 90 degrees to the beam illuminating it the degree of polarization will be nearly 100 percent. But as you add more and more milk the maximum degree of polarization decreases. Light from an ordinary glass of milk is essentially unpolarized. To obtain 100 percent polarization all the light you receive must have been scattered through 90 degrees. But when you look at a thick collection of particles it is unavoidable that you receive light scattered through many angles, some less than, some greater than 90 degrees. That is, the light is composed of many beams, each with a different degree of polarization. If two or more partially polarized beams are superposed, the degree of polarization of the resultant cannot be greater than that of the most highly polarized component. Indeed, it is usually less, even zero. If you mix an apple with an orange you get fruit salad, not an apple; adding a banana improves the flavor but takes you further from an apple. Similarly, multiple scattering tends to reduce the degree of polarization to zero, even if the scatterers are capable of completely polarizing unpolarized light when acting singly. It is no wonder that cloudlight is unpolarized: clouds are multiple scatterers composed of droplets much larger than the wavelengths of visible light.

Sunlight is multiply scattered in our atmosphere, even when completely free of foreign particles. Although multiple scattering by atmospheric molecules is not large (compared with that by clouds, for example), it is enough to reduce the maximum degree of polarization of skylight to well below 94 percent. Multiple scattering not only reduces the maximum degree of polarization, it also alters the pattern of polarized skylight. If only single scattering prevailed there would be two *neutral points* in the sky, directions in which the light is unpolarized: directly toward the sun and directly away from it. Because of multiple scattering, however, there are three neutral points. When the sun is higher than about 20 degrees above the horizon there are neutral points within 20 degrees of the sun, the Babinet point above it, the Brewster point below it. These two points coincide when the sun is directly overhead and move apart as the sun descends. When the sun is lower than about 20 degrees, the Arago point is about 20 degrees above the anti-solar point, the direction opposite the sun. Arago, Babinet, and Brewster were among the pioneers in the study of the polarization of scattered light. It was Arago who discovered the partial polarization of skylight in 1811.

Solid and liquid particles in the atmosphere further reduce the degree of polarization from its theoretical maximum value in a pure molecular atmosphere. They increase multiple scattering and their size and shape are such that they usually do not completely polarize unpolarized light even when acting singly. On hazy or dusty days or in polluted regions skylight may not be perceptibly polarized no matter where or when you look.

The effect of the ground on the polarization of skylight is similar to that of multiple scattering. Reflecting ground is an indirect source of light. Multiple scattering aside, the light you receive from a given patch of sky is sunlight scattered through a single angle and groundlight scattered through many angles. Thus the degree of polarization can only be less than what it would be in the absence of groundlight.

Nonspherical molecules and foreign particles in the atmosphere, multiple scattering, and light from the ground all reduce the maximum degree of polarization at 90 degrees to the sun to less than 100 percent. Who claims the highest observed degree of polarization? I have never seen a value reported greater than about 85 percent. Perhaps some of my readers know what the world record is and where it was observed.

CELLOPHANE TAPE, FROST, AND AIRPLANE WINDOWS

Polarization of skylight leads to many more observable consequences than those I have discussed in the previous sections. To understand them requires laying a few foundations.

Physics, like politics, makes strange bedfellows. Cellophane tape, frost, and airplane windows may appear at first glance to have little in common. Although they all are transparent to visible light, more to the point of this chapter is their *birefringence:* the speed with which light propagates through them depends on its polarization state.

That light can be polarized is demonstrated easily enough with polarizing sunglasses if the lenses can be popped out of their frames. If you look at light from a lamp through *one* lens while rotating it, the intensity does not change. But with *two* lenses, one on top of the other, the intensity varies as the lenses are rotated relative to one another. If the intensity is greatest for a particular orientation then it will be least when one lens is rotated by 90 degrees, and in this orientation the lenses are said to be *crossed*. Evidently, the first lens somehow modifies the light from the lamp in a way that can be detected by the second lens but not by our eyes.

Associated with a polarizing lens is an axis, which we cannot see, called its transmission axis. Light polarized with its electric field parallel to this axis is transmitted without loss of intensity. Very little light is transmitted if the field is perpendicular to this axis. The lens is a polarizing filter: it filters light according to its polarization state rather than its wavelength. If unpolarized light is incident on a lens, one component is transmitted, the other is strongly absorbed by the lens; thus, the emerging light is linearly polarized along the transmission axis.

Polarizing sunglasses are linearly *dichroic:* they absorb light differently depending on the direction of the electric field. Sunglasses and cellophane tape (not "transparent" tape that is only translucent) can be combined with a flat piece of clear glass and an overhead projector in a colorful demonstration of (linear) birefringence: the speed of light in the tape is polarization dependent. Lay a criss-cross of strips of tape, several layers thick, on the glass and place it between two crossed lenses. A subtly beautiful mosaic, suggestive of a stained glass window, will be projected onto the screen.

Although cellophane tape cannot polarize unpolarized light (look at the tape-coated glass through *one* lens while rotating it), it can change polarized light from one form to another (for example, linearly polarized in one direction to linearly polarized in a different direction). Viewed between crossed polarizers no light is seen if the glass is bare, but not if it is coated with birefringent tape. Light passing through the first lens is polarized along its transmission axis. Were it not for the tape most of this light would be absorbed by the second lens. Instead, some of it is transmitted; the amount depends on the thickness of the tape, its orientation, and the wavelength of the light. We see a mosaic of colored patches because the projector is a source of light of all colors.

White frost on window panes is common enough, but have you ever seen colored frost? This is unlikely unless you were looking through a polarizing filter. Like cellophane tape, ice is birefringent. But where, you may ask, is the second polarizer, which we cannot do without? It is outside, everywhere outside: the atmosphere itself. The degree of partial polarization of skylight is greatest at 90 degrees to the direct sunlight, so it will not do you any good to look at the frost through a polarizer when the sun is shining through a window.

If you are impatient to observe birefringence in ice then it is not always necessary to wait for frost to form on your windows. I was once at a workshop in Snowbird, Utah. At lunch, basking in the sun on the Plaza, I was talking with Hardy Granberg, a geographer from McGill University, about polarized light and birefringence, common topics of conversation at ski resorts. A young ski instructor could not help overhearing. Finally, curiosity got the better of her. "Are you physicists?" she asked. "No, we just play at it," Hardy replied. But professors, wherever they may, be cannot help professing, and an eager audience, even of one, is likely to elicit an impromptu lecture. So away we went, one spelling the other. In the heat of our lecture we rushed to a nearby snowpack, scooped up some wet snow, and pressed it between two crossed polarizing lenses. Light reflected from the snow and observed through this sandwich gives the same kind of colored mosaic seen in frosted windows, although less vivid.

Ice is birefringent because of its crystallinity: its molecules are arranged in a regular manner, like seats in an auditorium. Glasses and plastics, like liquids, are amorphous, their molecules do not form an ordered array. As a consequence, they do not naturally exhibit birefringence. But it is possible to induce them to do so. By squeezing, for example. Or stretching, as was done to the cellophane tape during its manufacture. Birefringence requires the existence of special directions in a material, either natural or induced. Without aid you cannot see these directions, but polarized light can.

Plastic objects are usually not completely free of all stresses, especially near edges. Airplane windows are a good example. You can observe stress-induced birefringence in them with crossed polarizers. As with frosted windows, you need only one polarizer: the atmosphere provides the other. So take a pair of polarizing sunglasses with you on your next flight and look out the window.

You don't really even need polarizing sunglasses. On a long flight once I was lost in thought, dreamily dangling my reading glasses from my fingers. My reverie was broken suddenly when I became conscious of the lenses: the reflection of the airplane window in them was a beautiful splash of colors. All the ingredients were at hand: partially polarized skylight streaming through a birefringent airplane window. And the second polarizer? When light interacts with matter, be it in the form of gas molecules, small particles, or a smooth surface, it may become partially polarized. So ordinary lenses can polarize unpolarized light, which you can verify easily enough merely by looking at reflections in them through a polarizing filter while rotating it.

On a summer day not long ago I threw a stick—not just an ordinary one but a special training dummy, for which I had paid almost three dollars—into the Pecos River in New Mexico for my dog to retrieve. He jumped in excitedly, fighting his way through the turbulent waters to reach this stick, his favorite. Unfortunately, it began to ship water and disappeared just before he could get his teeth into it. There he was in the middle of the river, flailing about bewildered, looking for a stick that was on its way to the bottom. Without hesitating I dove in, fully clothed, and triumphantly rescued the stick from the cold waters. Only afterwards I discovered that during this misadventure I had lost my polarizing sunglasses. So we drove to Santa Fe to get another pair. I wouldn't think of going anywhere without polarizing sunglasses. They are indispensable windows onto an unexpectedly variegated world of colors and patterns.

20

Colors of the Sea

A sea of lead, a sky of slate
Arthur Symons

Why is the sea blue? The answer might very well be, What explanation would suit your fancy? Eminent authority can be found to support almost any explanation imaginable.

Toward the end of the last century the consensus was that pure water absorbs red light more than blue light. This selective absorption by water in conjunction with scattering by matter suspended in it was thought by John Tyndall, John Aitken, and the Belgian scientist Walthere Spring, among others, to give natural waters their hues. And they did not come to their conclusions lightly. Spring's investigations into the color of water were especially thorough and extended over many years. But in 1910, following a holiday at sea, Lord Rayleigh muddied the waters by asserting that the color of the sea is "simply the blue of the sky seen by reflection." Yet less than a dozen years later, C. V. Raman, another Nobel laureate, refuted Rayleigh's explanation and offered in its place the notion that selective scattering by the water molecules themselves is the cause of the sea's color. When giants like these disagree, what are we pygmies to believe?

Before trying to reconcile these seemingly contradictory explanations, each of which still has its adherents among respectable folks, I must state that there is no single cause for the colors of the sea. A clue to the multiplicity of causes is the variable appearance of the sea: passengers on a ship see the sea differently from passengers in an airplane looking down on it from 30,000 feet; coastal waters are different from mid-ocean waters; the sea may vary from season to season, day to day, even hour to hour; and the sea does not look the same in all directions. What you see depends not only on what you look at but when and from where as well. Only after each of these is specified can you weigh the relative importance of the various causes that combine to yield what is observed.

I shall not explain the colors of the sea by appealing to human authority but rather to observation and experiment illuminated by reason, the only ultimate authorities in science. These colors are best understood by considering each cause in isolation from the others.

ABSORPTION AND SCATTERING BY PURE WATER

Let us imagine earth without an atmosphere so that the sea is illuminated by direct sunlight only. We also take it to be free of all suspended matter and so deep that no light reaches its bottom. What would such a hypothetical sea look like?

Pure water absorbs red light more strongly than blue light, of this there is no doubt. This seems incompatible with our everyday experiences only because we usually view water with dimensions of tens of centimeters (e.g., a glass of water), whereas white light must be transmitted through a meter or more of water before it is appreciably colored by selective absorption. To convince yourself that the path length of light in an absorbing liquid can affect its observed color, merely add a *small* amount of food coloring to water in a container with unequal dimensions; the water will be most deeply colored when viewed through the longest path. Beer is another example. It is yellow only because of the dimensions of ordinary bottles. Beer in bottles of more heroic proportions would appear distinctly reddish.

Sunlight transmitted into our hypothetical sea would be a deeper blue the greater the depth because of selective absorption. But absorption by itself is not sufficient to give the sea a blue color when viewed *above* water. A mechanism is needed to return some of the light transmitted into the water to our eyes; selective absorption of this returned light will make it blue. We have agreed that the bottom is too deep to return light from the sea. What other possible mechanisms are there?

Tyndall thought that completely pure liquids and gases would be "optically empty," that is, when free of all suspended particles, they would scatter no light. But by the second decade of this century enough experimental evidence had been amassed to dispel any lingering doubts that even pure liquids and gases scatter light (see Chapter 13 for more on this). In particular, pure water scatters light. But it is easier to observe molecular scattering of light by benzene than by water: benzene scatters light about 15 times more than water. In a darkened room illuminate some benzene with a narrow beam of white light; the light it scatters will be bluish. Benzene and water, like the air molecules that make the sky blue, scatter blue light much more than red light. Indeed, the wavelength dependence of scattering by liquids and gases is about the same, although a given number of molecules in the gas phase scatters much more strongly than when in the liquid phase.

Raman was one of the pioneers in the molecular scattering of light. It is no wonder, then, that he sought the explanation for the color of the sea in this mechanism: If you go to a psychiatrist with a vague complaint, he is likely to find the source in your head; a proctologist, however, is likely to lower its center of gravity. Raman overstated his case by asserting that the blue color of the sea is really due to molecular scattering, "selective absorption only helping to make it a fuller hue." Suppose that there were no absorption. Then the deep sea would be *white,* just as a glass of milk is white even though it contains particles that scatter blue light more than red light (for more on this see Chapter 14). On the other hand, if there were no molecular scattering (and no other mechanism for returning light), the sea would be *black,* regardless of the absorptive properties of water. Both scattering and absorption by water, neither of which is more essential than the other, combine to give our hypothetical sea its color.

But what would it look like? How would we describe it? Without a doubt it would be deep blue of high purity, higher than that of the bluest sky, but of low brightness: molecular scattering is not sufficient to make water very bright, even when illuminated by bright sunlight. We would probably call such water blue-black, perhaps even black. To observe the deep blue color it would be necessary to photograph our sea for a long time. Something like this was done by Robert Greenler and reported in his splendid book *Rainbows, Halos, and Glories.* He photographed a moonlit sky with the shutter open for several hours. The resulting photograph is indistinguishable from that of a sunlit sky except for the streaked paths of stars.

The very essence of sea water is its saltiness. Does this affect its color? Although salt water scatters slightly more than fresh water, dissolved salts in the concentrations found in sea water do not appreciably change its optical properties.

What about dissolved impurities? My attitude toward impurities is perhaps extreme: they should be invoked only after all else fails. Saying that such and such is caused by impurities is often no more than saying that I am ignorant of its cause and I sweep my ignorance under a rug labeled "impurities." My attitude is shared by John Aitken, who wrote these words in 1882:

> The cause of the color of water has been a frequent subject of speculation. Every substance which has been discovered in water has in turn been suggested as the cause of the color. When no useful purpose could be given for its presence, it was told off to do the ornamental, and make the water beautiful to the eye. All these speculations assume that the color of water is due to some impurity in the water. This, however, is obviously begging the question. It is first necessary to find out whether water has any color in itself, and what that color is, before we can say anything about the effect of impurities.

Water in itself is blue, it needs no help from dissolved coloring matter. Such matter exists in the sea, especially near coasts, but it is highly variable and localized. It may modify the appearance of the sea, yellow matter, for example, shifting the color from blue to green. It is one spice among several that relieves the sea of its inherent monotonous blueness.

Before discussing the effect of suspended matter on the colors of the sea, it is well to briefly mention another mechanism inherent in the sea itself for returning light from its depths. Molecular scattering by water may be looked upon as arising from *microscopic* fluctuations in density, that is, pure water is homogeneous on average, but there are *local* variations from this average. Because of temperature and salinity gradients in the sea its density may not be uniform even on a *macroscopic* scale, and this may also direct light from its depths to the surface. Simple demonstrations to illustrate this are easy to devise. For example, a beam of light directed above a hot plate onto a screen does not give a uniformly bright spot but rather a flickering pattern of light and dark. Light that would have gotten to the screen in the absence of heating is deflected because of the density variations induced by heating. Of course, in this demonstration the temperature gradients are much greater than those found in the sea, but this is a difference of degree not of kind. Where there are strong density gradients in the sea, they may affect its brightness and color.

SUSPENDED MATTER IN SEA WATER

Anyone who has ever tried to prepare clean water for sensitive light scattering experiments knows that it is a task to try the patience of a saint. To such an experimenter, even ordinary distilled water is intolerably filthy. Thus it is reasonable to expect sea water to contain suspended matter. Because of their nearness to land, coastal waters are sometimes quite turbid, and the global winds disperse particles even to the remotest oceans. Upwelling is another mechanism for dispersing particles in the sea. It does not take many such particles before scattering by them overwhelms that by molecules. This was evident in one of the demonstrations described in Chapter 14. A few drops of milk added to an aquarium filled with clean water markedly increased scattering. In all but the cleanest waters, therefore, molecular scattering will be only a small part of total scattering. Raman's explanation of the blue sea was correct to the extent that molecular scattering contributes to what is observed, but it is a minor contributor. Without it, I believe, we would still have a blue sea.

It is less important how particles in the sea scatter light than that they merely do so. Much the same results are obtained with particles having vastly different scattering properties. To demonstrate this I followed some suggestions made by Aitken in 1882, although not without adding a few tricks of my own.

I painted several jars black and filled them with water to which blue food coloring had been added. But the blue color was barely evident until some particles were added. I used milk, the particles in which scatter blue light preferentially, and precipitated chalk (calcium carbonate), which scatters visible light of all wavelengths about equally. Depending on how many particles were added, various colors were obtained. Indeed, I could obtain almost the same color with milk or chalk. The variations on this demonstration are many. Try your own, don't take my word for it—or for anything else.

Adding particles to the water increases its brightness at the expense of its spectral purity. The greater the concentration of particles the shorter the average distance between scattering events. More incident photons escape absorption by the water and re-emerge from it without having traveled paths long enough to develop its full blue color. The result is a shift from the blue toward the green, brighter but of lower purity.

To forestall possible objections that my demonstration is unrealistic because the concentration of particles may be much greater than that usually found in the sea, I point out that absorption by the blue water in my jars is also much greater than that by sea water. It is the *ratio* of scattering to absorption that determines the brightness and color of an optically thick multiple-scattering medium, not the values of these two quantities separately. Thus to simulate the sea with a more strongly absorbing liquid, scattering must be proportionately increased.

For this demonstration I purposely chose colorless particles. It is evident from the demonstration that colored particles are not necessary to give water blue-green colors. Neither plankton—which comes in various colors, by the way—nor algae nor anything other than white particles are necessary to give what is observed. My experience with rocks on this planet is that if they have any color at all it is usually yellow or red. Particles in the sea originating from such rocks will enhance the greenness of sea light. In sufficient numbers they may make the sea yellow or even brown.

Just as liquid droplets in a gas scatter light, so also do gas bubbles in a liquid. There are air bubbles in the sea; you can learn about them in Duncan Blanchard's book, which I mentioned in Chapter 2. The effect of bubbles on the color of the sea may be observed in breaking waves. Such waves are endlessly fascinating; I have watched hundreds of them. Although each is different, they share common features. Where a great many air bubbles have been entrained by a breaking wave it is white. But where there are fewer of them it is blue-green or green, brighter than the sea but not as bright as the foamiest parts of the wave. Even after a wave has broken and the water is again quiescent, a pastel green patch often remains, slowly fading into the surrounding sea as the bubbles dissipate. Thus the effect of bubbles on the color of the sea is similar to that of solid particles.

I look upon the effect of the sea's bottom on its color as merely a variation on the suspended particle theme. That is, the bottom is merely particles concentrated at a single depth instead of distributed more or less uniformly throughout the water. In sufficiently shallow water, tens of fathoms or less, the bottom may make its presence apparent. Like particles it can be white or colored. My observations of bottoms—of the sea, that is—is that they tend to make the sea brighter and greener.

When you fly over coastlines be alert for sudden changes in the color of the sea. The deep blue of the open sea often gives way to a brighter blue-green or green. Many factors contribute to this color change, and I would be unwilling to say which of them is dominant in any instance without careful observations and possibly even experiments. But I hope that by now it is clear that several factors can combine to yield what is observed: a greater number of suspended particles of all kinds, the shallow bottom, even breaking waves and, for all I know, submarines, great white sharks, and sunken Spanish galleons—anything to shorten the paths taken by photons on their way out of the sea.

RESTORATION OF THE ATMOSPHERE

Up to this point I have considered a hypothetical sea on a planet without an atmosphere. To explore further the appearance of the sea I must restore the atmosphere to its proper place. On the real planet earth, as opposed to a hypothetical one, both light from the sky and direct sunlight illuminate the sea. But light from the sea derived from the sky does not destroy that derived from direct sunlight: the two combine to give what is observed. My task is not to declare one or the other the cause of the sea's blueness but to determine their relative contributions under various conditions.

Before doing so I should point out that color is not a property of objects but rather of light from them. An object can have quite different colors depending on its illumination. It is therefore misleading to talk about *the* color of an object as if it were one of its attributes. I use phrases like "colors of the sea" merely as shorthand for colors of the light from different patches of the sea under various conditions of illumination and observation.

COLOR AND BRIGHTNESS VARIATIONS OF THE SKY

A student of mine was once told that the blue sky is the reflection of the blue sea. Aside from doing violence to conservation of energy—the sky is generally brighter than the sea—this explanation would not travel well: folks in Kansas, for example, might have difficulty accepting it. It is more usual to be told that the blue of the sea is reflected blue light from the sky. Yet a few observations shake one's belief in the completeness of this explanation. For example, I have seen colored seas on days when there was enough thin cloud cover to whiten the sky but not to greatly reduce the intensity of illumination. On more overcast days the sea may have a leaden appearance, which has been adduced as evidence to support the blue sky reflection theory. The argument goes something like this: on overcast days when the sky is not blue the sea, too, is not blue, hence the blue sky is the cause of the sea's blueness. Merely because two events—blue sky and blue sea—occur together does not necessarily imply that one is the cause of the other. Both may have a common cause, sunlight in this instance.

To fully understand the role of the sky in determining the colors of the sea, we must know more about the sky than that it is merely blue. How blue is it, for example? Previously, I have used the term *purity* of color somewhat loosely, hoping that my meaning would be clear from the context. The time has come to be more precise.

We say that the sky is blue, but this does not mean that all colors except blue—itself an imprecise term—are absent from skylight. To the contrary, it is composed of all colors, although blue dominates (see Chapter 12). The extent to which it dominates is the light's purity. We may treat light from any object as if it were a mixture of white light and light of a single *wavelength,* the *dominant wavelength,* the relative contribution of which to the total light determines its purity. Under ideal conditions (e.g., no multiple scattering in an atmosphere free of particles), the sky has a purity of about 41 percent. Real blue skies will have a purity less than this.

How pure is light from the sea compared with that from the sky? The sea often appears to be a purer blue than the sky, which is evidence against the reflection theory. But we must be careful in making this inference because the sea is not as bright as the sky and we might be confusing brightness differences with color differences. The only way to make a fair comparison is to observe the sky and the sea under conditions of equal brightness. Although I have not done so, this might be accomplished by comparing the sea with the sky reflected by black glass. Let us assume, however, that the sea is a purer blue than the sky. Lord Rayleigh had a ready response to this: "we are apt to make comparison with that part of the sky which lies near the horizon, whereas the best blue comes from near the zenith." Rayleigh's assertion is indeed correct. Not only is the sky not pure blue, its color and brightness are not uniform. It is evident from Figure 20.1 that even if we ignore the clouds near the horizon, the sky over the sea is brighter toward the horizon. This, as we shall see, is relevant to the

Figure 20.1 Atlantic Coast, Massachusetts. Note the nonuniform brightness of the sky.

Figure 20.2 Even in the extremely clear air of Southeast Iceland the sky is brighter toward the horizon.

appearance of the sea. Before demonstrating this, I must digress briefly on why the horizon sky is brighter and less pure than the zenith sky.

The white band of the sky near the horizon is often blamed on pollution. This is one of those fallacies that has become accepted because of frequent repetition. The best way to dispel it is to make a few observations.

The photograph for Figure 20.2 was taken in a remote part of Iceland not long after a heavy rainfall. Even if I had not remembered this, it would have been safe to say that *any* photograph from Iceland was taken not long after rain. Despite its name, Iceland is not an especially cold country (if you crave really low temperatures, try more southerly places such as Montana or the prairie provinces in Canada), but it is wet and windy. The population of Iceland is less than a quarter million; its area is about that of Pennsylvania. Approximately half of Iceland's population lives in Reykjavik, where natural hot water is used for domestic heating. There is very little heavy industry; the nearest "dark, Satanic mills" are in Glasgow, about a thousand miles downwind. For several reasons the air in Iceland is very clean. It is the cleanest air I have ever gazed through. As I pointed out in Chapter 16, I sometimes have greatly underestimated distances to mountains because of the unaccustomed visibility. Yet Figure 20.2 shows a white horizon where it should not be according to the conventional wisdom. And I have seen photographs taken in the Antarctic showing white horizons on cloudless days. If pollution were necessary then the horizon could not be white. Why, then, is it white?

Pollution only enhances what is inevitable. Even if our atmosphere were completely free of particles, the horizon would still be white (I assume that the sun is not near the horizon). In a given direction skylight is light scattered by all the molecules and particles along the path in that direction. Incident sunlight is composed of all wavelengths; the shorter ones are more likely to be scattered toward your eye, but the longer ones are more likely to make it to your eye without being scattered again. There are fewer scatterers along a path through the atmosphere toward the zenith than toward the horizon (Fig. 20.3). Hence, along a zenith path the scattered sunlight, which is enriched in shorter wavelengths, is likely to reach your eye without being scattered again. Near the horizon, path lengths are longer. Hence, the horizon sky is brighter at the expense of purity of color.

There is a natural tendency to think that if a little bit of scattering makes the sky blue then a lot of scattering will make it even bluer. But this is incorrect: single scattering giveth, multiple scattering taketh away. If our atmosphere were perhaps ten times more massive, the sky would be white everywhere. If, on the other hand, it were a tenth as massive, the sky would be black overhead and bluer toward the horizon.

Now that we better understand why the color and brightness of the sky are not uniform, we are ready to return to the appearance of the sea.

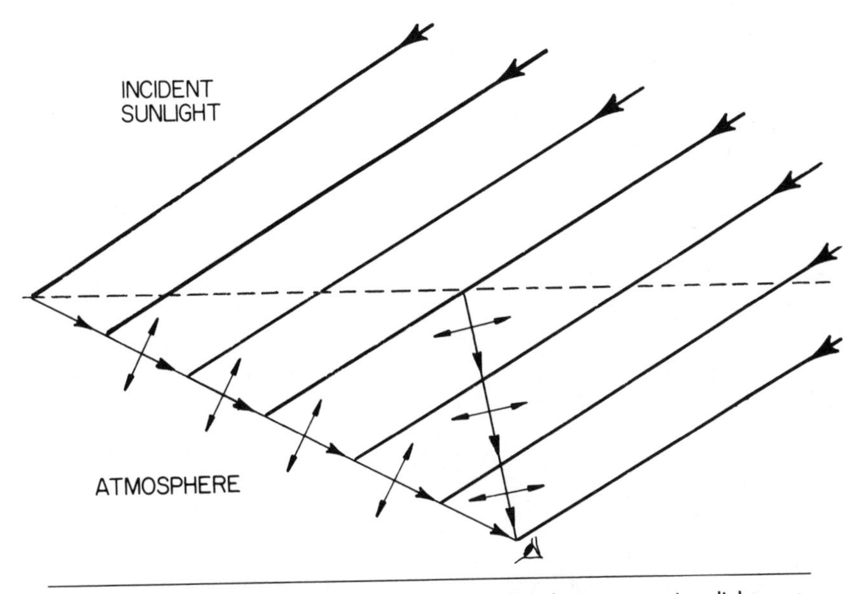

Figure 20.3 Path lengths in the atmosphere. An observer receives light scattered by all the molecules and particles along the line of sight. Paths near the horizon are longer than those near the zenith, hence the horizon sky is brighter. From *The Physics Teacher*, C. F. Bohren and A. B. Fraser, May 1985.

LIGHT FROM A FLAT SEA

The sea is rarely—if ever—perfectly flat. But let us pretend that it is. We can add ruffles when the mood strikes us. An observer looking at a particular patch of the sea receives reflected skylight (I assume that the sun is behind him so that he receives no reflected direct sunlight, or glitter) and upwelling light from the sea. The relative contribution of each depends on the brightness reduction of skylight upon reflection by the air-water interface and the brightness reduction of upwelling light upon transmission by the water-air interface. These, in turn, depend on the *observation angle*, the angle between the observer's line of sight and the perpendicular to the sea. The dependence of brightness reduction on observation angle is shown in Figure 20.4.

Note that water is weakly reflecting except at large observation angles; at 90 degrees (glancing incidence) it is a perfect mirror. You can verify this with a blackened pie pan filled with water. Look at the reflection of an object in the water at different observation angles, large angles in particular. A piece of black glass also serves to show how reflection increases with increasing angle.

Over the range of angles for which incident light is greatly reduced in brightness by reflection, it is reduced much less by transmission. Even if incident light were completely transmitted, however, there would still be a *brightness* reduction of about 50 percent: because of refraction the transmitted light is less concentrated than the incident light, hence less bright. You can observe this

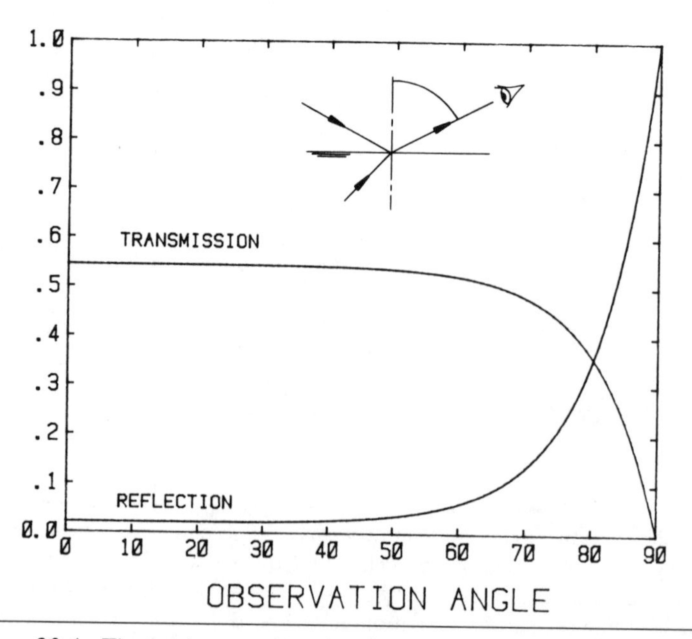

Figure 20.4 The brightness of incident light is reduced upon reflection and transmission by water.

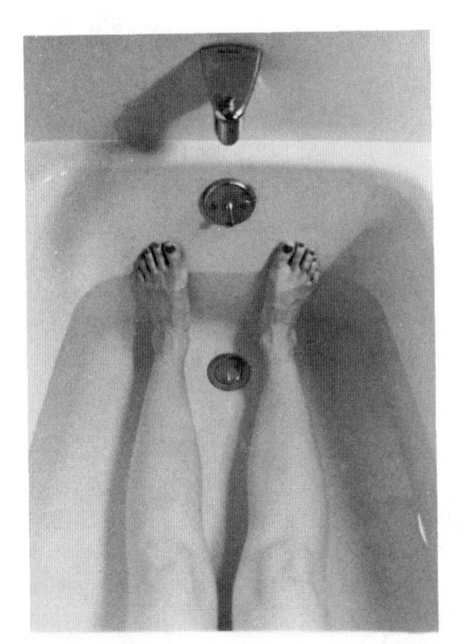

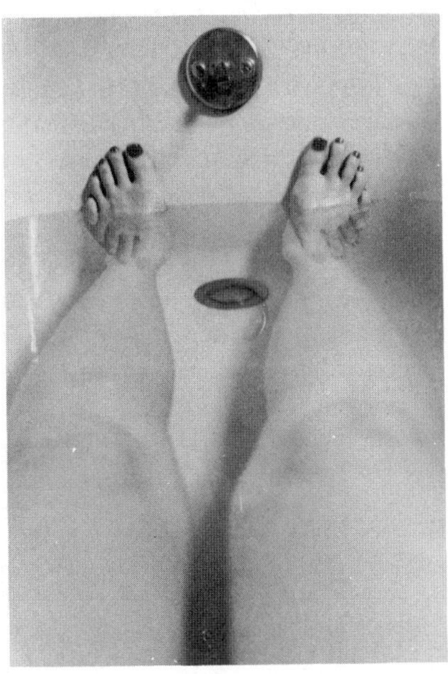

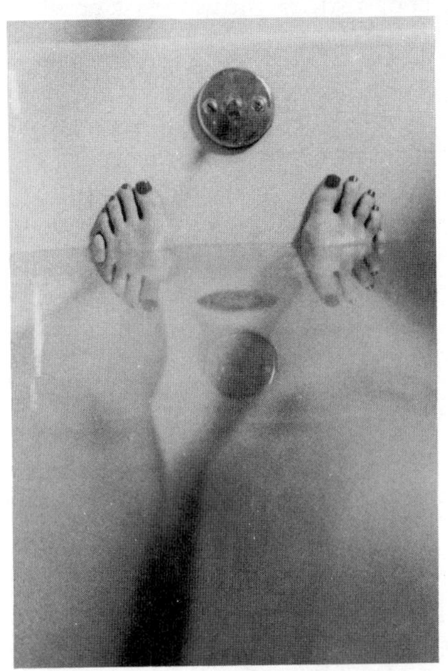

Figure 20.5 The qualitative correctness of the curves in Figure 20.4 can be verified in the bathtub. When looking nearly perpendicularly to the water, the drain, seen by transmitted light, is much brighter than the reflected image of the shiny plate on the side of the tub (top). But as the angle of observation becomes more oblique, the relative brightness of these two changes (middle), until at near-glancing incidence, they are almost equal (bottom). Note also how the brightness of the images of feet under and above water changes with angle.

brightness decrease by putting an object—white plastic spoons work well—in water in a blackened pan and comparing it with the same object above water. At the same time you can compare it with its much less bright reflected image.

Observations can also be made in bathtubs. If you don't take baths, you should. Great discoveries have been made in bathtubs—only bad singing issues from showers (see also Chapter 7). While lazing in the tub look at your feet under water and the reflected image above them of something on the side of the tub, a drain or faucet perhaps. As you move your eyes closer to the water, the reflected image will increase in brightness while that of the transmitted light will decrease. If you don't mind getting your hair wet, you can bring your eyes right down to the water's surface. Note carefully what you see when you do (Fig. 20.5).

Those with an aversion to water can perform a drier experiment, one with stronger links to the question at hand. Fill a blackened pan to its brim with water to which blue food coloring has been added. Then add some particles (e.g., milk), enough to make the water bright blue. With a white backdrop behind the pan, look at the water from various angles. Viewed perpendicular to its surface it will be blue, but as you move your line of sight closer to the horizontal the blue will become less intense. At about five degrees above the horizontal (the precise angle depends on the number of particles) the water will be white: reflected light from the backdrop (i.e., the local horizon) will dominate trans-

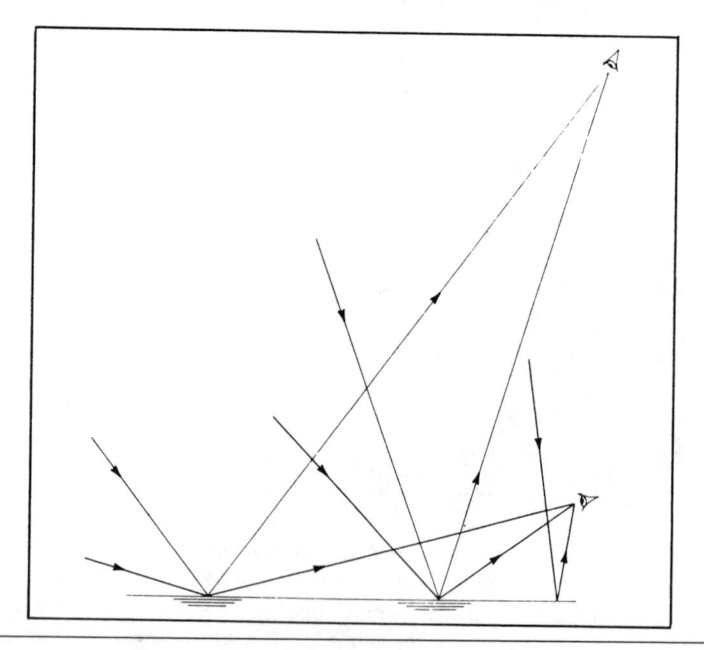

Figure 20.6 Reflection by a flat sea. Depending on an observer's location and the part of the sea viewed, the reflected skylight received comes from parts of the sky that vary both in color and brightness.

mitted blue light from the water. This further confirms the correctness of the curves in Figure 20.4, which shows that reflection dominates transmission for observation angles greater than about 85 degrees.

Now consider an observer at the edge of a flat sea (Fig. 20.6). The observation angle for that part of the sea more than about 60 feet from him will be greater than 85 degrees. For these angles, reflected skylight will dominate transmitted light from the sea, particularly if the brightness of skylight is greater than that of upwelling light from the sea. But this skylight comes from low in the sky (less than five degrees above the horizon), where it is whiter and brighter than overhead. For this observer, the color of most of the sea before him is indeed that of the sky, but white rather than blue.

An observer in an airplane will not agree with his counterpart on the shore about the color and brightness of the *same* patch of sea. Observation angles for the airplane passenger are generally smaller; as a consequence, he sees a darker, bluer sea, not only because transmitted blue light from the sea is brighter (it may dominate if there is enough suspended matter), but also because the reflected skylight comes from closer to the zenith. This underscores my previous assertion that the color of an object need not be one of its unique attributes; if it were, both observers would agree on the sea's color, but they do not.

LIGHT FROM A RUFFLED SEA

The sea is never still. Waves of all sizes, from long transtidal waves to short capillary waves, continually disturb its surface. How do they affect light from the sea?

Suppose that an observer is situated so that he sees both sides of waves. Note in Figure 20.7 that the front of a wave is tilted *toward* him. Thus he receives light from higher in the sky, a darker and purer blue but with a greater reduction in brightness upon reflection, than he would if the sea were flat. The back of a wave is tilted *away* from him, so he receives light from lower in the sky, brighter and less pure but with less reduction in brightness upon reflection, than he would if the sea were flat. This pattern of light and dark can be seen in the photograph in Figure 20.8. Note also how the brightness of the sea varies with distance from the shore. Near shore, there is a definite pattern of alternating dark and bright narrow bands. Farther from shore, there is a bright broad band, light from clouds reflected by a fairly calm sea. Farther still, the sea abruptly becomes darker where the sea is rougher; in this region the distance and wave slopes are such that you can see only the fronts of waves.

The sea mirrors a sky of nonuniform brightness and color, but what part and to what extent depend very much on the state of the sea and the position of the observer. At the same time, upwelling light from the sea is of greater or lesser importance depending on these same factors. Is it any wonder, then, that the colors of the sea cannot be treated glibly in a sentence or two? The world was not designed for the convenience of those who frame multiple-choice examinations.

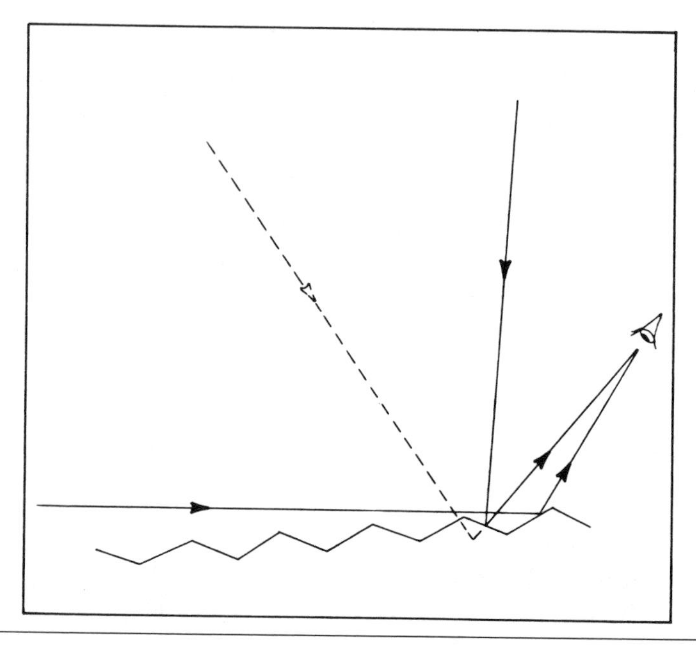

Figure 20.7 Reflection by a wavy sea. The front of a wave reflects light from higher in the sky and the back reflects from lower in the sky than a flat sea would.

Figure 20.8 Note the pattern of light and dark on the water. If you look carefully you can see the reflections of clouds.

AIRLIGHT

There is another source of light which, to my knowledge, has never been mentioned in the context of light from the sea: airlight (see Chapter 16).

When we look at the sea we unavoidably see airlight, light scattered by all the molecules and particles along the line of sight. When you are standing on a high bluff overlooking the sea, between you and most of it there are miles and miles of air at its densest. Light scattered by this air adds to that reflected and transmitted by the sea. Even if the sea did not direct any sunlight or skylight to our eyes, it still would not appear uniformly black: close by it would be black, bluer farther away, and in the far distance indistinguishable from the white horizon, just like the dark ridges in Figure 16.2.

ELIMINATING REFLECTED SKYLIGHT

To determine the relative contributions of reflected skylight and transmitted upwelling sea light to what we observe, it would be best to eliminate one or the other. This is less impracticable than it might seem at first glance. Reflected light could be eliminated by, for example, observing the sea from a glass-bottomed boat. There is another way: using polarizing sunglasses.

I pointed out in the previous chapter that we may consider any light to be a mixture of completely polarized light and unpolarized light. The degree of polarization is the relative contribution of the polarized component to the total. Skylight is partially polarized, the degree of which is greatest in directions perpendicular to the sun.

Reflection by smooth surfaces is a mechanism for polarizing unpolarized light (if it were not, it would be pointless to buy polarizing sunglasses). This is also easy to observe: just look at light reflected by smooth objects through polarizing sunglasses while rotating them. The degree of polarization acquired by unpolarized light upon reflection by water exceeds 50 percent for observation angles (angles of incidence) between about 30 and 70 degrees.

For two reasons—skylight is partially polarized and reflection polarizes—skylight reflected by water may be highly polarized. Transmitted light from the sea, however, is weakly polarized. It is therefore possible to eliminate more reflected light than transmitted light with a polarizing filter. To obtain the highest degree of polarization requires a judicious choice of time and place. You would want to be well above the sea, not on the shore, with the sun at your back, perhaps 30 to 60 degrees above the horizon.

What you might see under these conditions is that as you rotate the filter, the color and brightness change. I have observed this on a bluff about 100 feet above the sea. For this coastal water, and from my vantage point, the effect of skylight was to make the sea bluer and brighter than it otherwise would have been. Without skylight the sea was much greener. I never have had the opportunity to observe the sea far from shore (except from an airplane). If you get such an opportunity be sure to record all details—sun angle, sky condition, sea state, your elevation—so that your photographs can be interpreted properly.

HISTORICAL POSTSCRIPT

Light from the sea is a mosaic. To this extent almost every explanation of the sea's color is correct—but incomplete. We owe to Lord Rayleigh the notion that "the much admired dark blue of the deep sea has nothing to do with the colour of the sea, but is simply the blue of the sky seen by reflection." Yet in each paragraph following this unequivocal assertion at the beginning of his 1910 paper, he adds more and more qualifiers. Even Rayleigh did not fully accept his own explanation; he was too brilliant a scientist not to be aware of the several causes that combine to give what is observed. His final about-face is a postscript in which he notes that "the colour of the Mediterranean and other waters was long ago [1882] discussed by Mr. J. Aitken—an excellent observer. . . . His principal conclusions are very similar to my own." With all due respect to the memory of Lord Rayleigh, I cannot agree. In only one paragraph out of dozens does Aitken even mention reflected skylight, and then almost parenthetically: "The effect of the light reflected from the surface of the water is also of importance." Compare this with Rayleigh's unequivocal assertion about reflected light.

Lord Rayleigh explained the blue color of the sky, one of his many scientific triumphs. Did he unconsciously—and against his better judgment—wish to make a clean sweep of blueness in nature by reducing the color of the sea to a problem he had already solved? This is merely speculation, of course. Beyond speculation, however, is the fact that Rayleigh and Aitken arrived at quite different conclusions about the role of reflected light in giving color to the sea, despite Rayleigh's statement to the contrary.

The struggles even these two giants had with the colors of the sea underscore a recurring theme of this chapter: a satisfactory explanation does not come easily.

21 Indoor Rainbows

The rainbow . . . does not attract an attention
proportionate to its singularity and beauty.
Moses was the last to comment on it.
Henry David Thoreau

Are poets more moved by rainbows than scientists are? Does too much knowledge dull one's appreciation for beauty? Thomas Campbell certainly must have thought so when he wrote these lines:

> Triumphal arch, that fill'st the sky
> When storms prepare to part,
> I ask not proud philosophy
> To teach me what thou art

Perhaps he was right in disdaining what "proud philosophy" could have taught him about the rainbow. But he might have had second thoughts if he had been with me at the University of Arizona one day during a rainshower. At the time I was working in my office in the Institute of Atmospheric Physics. Suddenly, people were rushing upstairs to a room with a commanding view of the campus. Like a sheep, I followed the throng. A rainbow had been spotted, and it was a beauty. I can close my eyes and still see its stunning supernumerary bows. And I also remember the people who excitedly shared this visual delight with me. Some of them were experts on light scattering; they knew the physics of the rainbow inside and out. They could fill blackboards with equations describing it. And yet when confronted with one they were like children, their eyes agleam at seeing what they had undoubtedly seen many times before and understood in minute detail. Proud philosophy had not dulled their senses. If anything, it had made them sharper, which is what I hope to do for you in this chapter.

THE RAINBOW: THE ESSENTIAL INGREDIENTS

What causes the rainbow? Briefly, it is scattering of sunlight by raindrops. They scatter light more strongly in some directions than in others; and in any given direction, they scatter light of some wavelengths more than others. Although this explanation is correct, it is not very satisfying. It doesn't provide much meat for the mind to chew on. It is more a menu than a meal.

How one chooses to describe the rainbow is a matter of taste and experience. There is a hierarchy of theories, successive members of which are ever more faithful to reality at the expense of greater complexity. Moreover, these theories may little resemble one another. In Chapter 18 I discussed the forward-backward asymmetry in scattering by large particles by considering them to be arrays of dipoles; I did not invoke, even implicitly, concepts such as rays and the laws of reflection and refraction. I could take the same approach to discussing the rainbow, but the results would hardly be worth the stupendous effort. Fortunately, much simpler approaches are adequate. Geometrical optics, the optics of rays, describes some, but by no means all, of the features of rainbows. But this does not mean that geometrical optics causes the rainbow. No theory causes anything.

Strictly speaking, there is no such thing as a ray of light, that is, a beam with vanishingly small transverse dimensions. All attempts to isolate a ray fail: if one tries to narrow a beam by passing it through smaller and smaller holes it will

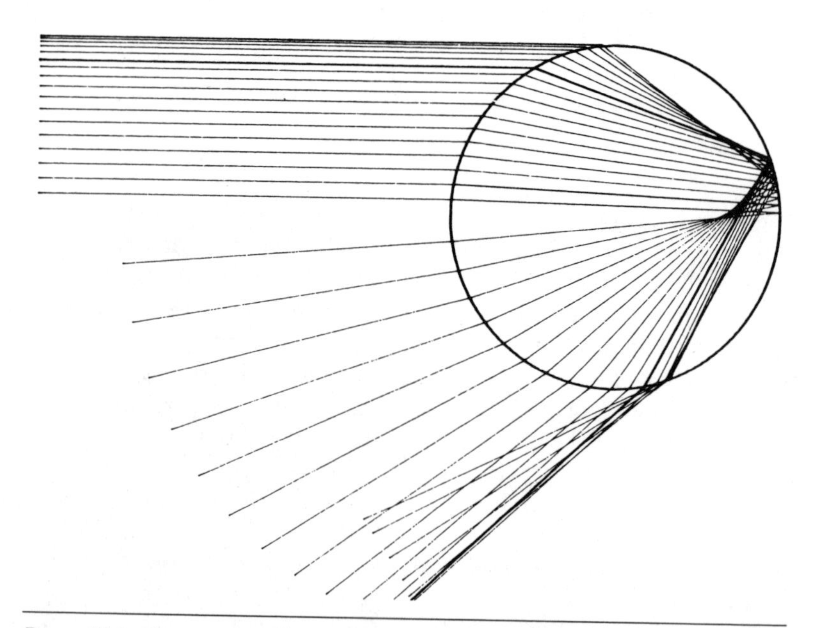

Figure 21.1 Rays incident on this sphere are refracted upon entering the drop at the first interface, then reflected internally, and finally exit after another refraction. Note that although the incident rays are uniformly spaced along the surface, those that exit are not. There is an angle of minimum deviation around which exiting rays cluster; this angle is the rainbow angle, at which, according to geometrical optics, the intensity is infinite. Figure courtesy of Alistair Fraser.

become broader rather than narrower when the holes are comparable with the wavelength of light. Nevertheless, a ray of light is a useful idealization—if you keep in mind its limitations.

Consider now a single raindrop illuminated by a beam with transverse dimensions larger than the drop diameter. Because the drop is much larger than the wavelengths of visible light we may pretend that the beam is composed of a collection of many rays and trace the fate of each incident ray by applying the laws of reflection and refraction at each interface (Fig. 21.1). Although the incident rays are uniformly spaced along the surface, those exiting are not. They concentrate near the rainbow angle, which is why the rainbow is bright.

A laser beam can be a good approximation to a ray of light, and a realization of the ray diagram in Figure 21.1 can be obtained by illuminating a water-filled beaker with a laser beam. If a small amount of milk is added to the water, the fate of the beam after it enters the water is more evident. The incoming and exiting beams can be made visible by spraying talcum powder along their paths, as was done when the photograph for Figure 21.2 was taken.

Although a beaker is not spherical, its cross section (perpendicular to its axis) is circular, as is that of a spherical water drop. Thus the various angles of reflection and refraction for a water-filled beaker are the same as those for a spherical water drop.

To show that there is an angle of minimum deviation (the primary rainbow angle) of the beam that exits after one internal reflection merely illuminate the beaker at various angles of incidence. Note how the deviation angle of the ex-

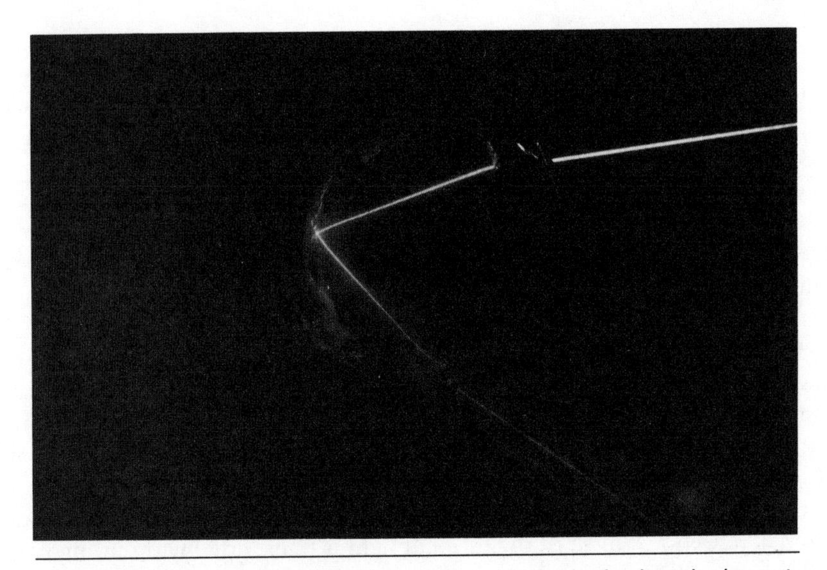

Figure 21.2 Ray tracing with a laser beam. Inside the beaker the beam is visible because milk was added to the water. The incident and exiting beams were made visible by spraying talcum powder along their paths.

iting beam—the (greatest) angle it makes with the incident beam—changes as the angle of incidence changes; this will be more apparent if you project the exiting beam onto a screen. All the incident beams must be parallel to one another, which can be achieved by fixing the laser in place and moving the beaker along a line perpendicular to the beam. Move the beaker so that the beam makes all angles of incidence with the beaker, from zero degrees (normal incidence) to 90 degrees (glancing incidence). As the angle of incidence is increased, the deviation angle of the exiting beam will first decrease uniformly then increase uniformly (see Fig. 21.1). This shows that there is an angle of incidence such that the deviation angle of the exiting beam is a minimum; this is the primary rainbow angle.

Light internally reflected more than once before exiting also gives rise to rainbow angles, an infinity of them. Each successive bow, however, is less bright than its predecessors. Despite travelers' tales to the contrary, natural rainbows of orders higher than the second are rare—I have yet to see a photograph of one—because they fade into the background illumination. Recently, however, what appears to be a reliable observation of a tertiary rainbow was reported by D. E. Pledgley (*Weather*, December 1986, p. 401). Unlike the primary and secondary bows, the tertiary bow is seen looking toward the sun, about 40 degrees from it. Bows of even higher order may be observed in the laboratory, where background illumination can be eliminated. To learn more about them I highly recommend Jearl Walker's article in *American Journal of Physics* (1976, Vol. 44, p. 421).

AN INDOOR RAINBOW

The laser demonstration described in the previous section shows that at special angles—rainbow angles—light scattered by a water drop is much more intense than at other angles. But a rainbow is more than just an intense arc of light. Because the angle of minimum deviation depends on wavelength, light of different colors is separated in natural rainbows. Indeed, "rainbow" evokes images of a profusion of colors. A polychromatic source of light is therefore required for a rainbow—as that word is commonly understood—rather than a nearly monochromatic source such as a laser.

Glass globes filled with water and illuminated by slide projectors (or sunlight) have been used to demonstrate rainbows. In my experience the rainbows that result are adequate but not very impressive (for a different viewpoint, see the letter by Arvid Skartveit and Frank Cleveland in the October 1985 issue of *Weatherwise*). This is because only a small fraction of the incident light contributes to the various rainbows, which are formed by light that undergoes one or more internal reflections; most of the incident light is transmitted without internal reflection.

A colleague of mine at the University of Arizona, Donald Huffman, devised a clever way of greatly enhancing the brightness of the rainbow obtained with a water-filled globe: he deposited, in a vacuum, a mirror coating on part of the globe. The rainbow angles are not changed, but the brightness at these angles is greatly increased: light that would have been lost to the rainbow is directed into it by the mirror.

Although Huffman's variation on a common rainbow demonstration is a great improvement, it requires apparatus that is not readily available. In a discussion about this with Dennis Thomson, one of my colleagues at Penn State, we hit on the idea of wrapping a flexible mirror around a beaker. The mirror was a sheet of aluminized plastic film (not aluminum foil, which is likely to be too crinkly). The beaker shown in Figure 21.3 has such a mirror covering part of its surface; this is why you see no light transmitted through the beaker. I also tried painting part of the beaker with aluminized paint; the results were less satisfactory, although someone who takes more care might have more success than I did. In any event, aluminized film is almost as common as aluminized paint.

You need not cover an entire half of the beaker with aluminized film. The ray diagram in Figure 21.1 will guide you in determining what fraction of the beaker's surface needs a mirror. For example, if you are interested in angles near the rainbow angle, only a small amount of the surface needs a mirror.

Figure 21.3 shows what is obtained using a water-filled beaker with a mirror covering part of its surface. The source of illumination is a slide projector in which an opaque slide with a slit in it is inserted. Note how bright it is inside the bow. This illustrates a point Alistair Fraser often makes: the rainbow is merely the bright edge of a disc of light.

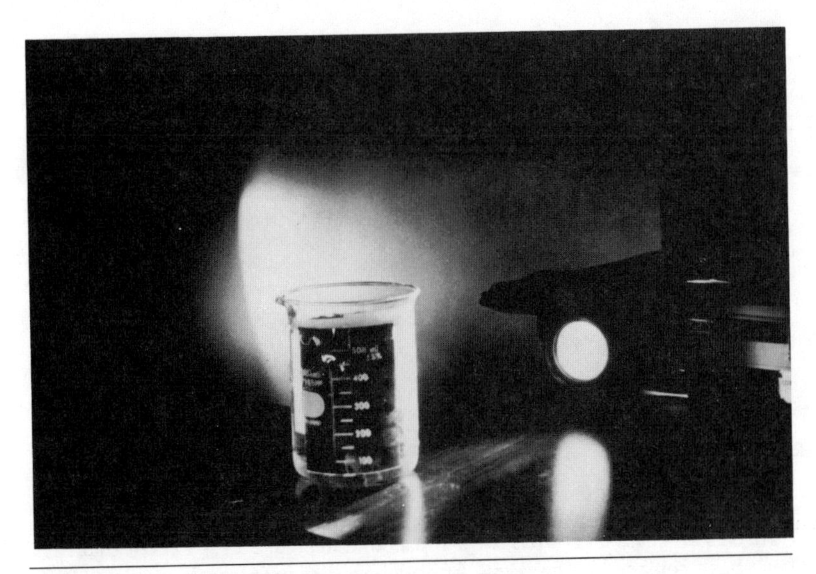

Figure 21.3 An indoor rainbow—without rain.

SUPERNUMERARY BOWS

Geometrical optics applied to large water drops predicts the existence of rainbows, their angular position, and their color separation. But it fails completely to account for supernumerary bows, even though Nature has no trouble making them. Here is one example of the inadequacy of geometrical optics. This is hardly cause for hand-wringing. Geometrical optics is so simple that it is somewhat of a surprise that it works at all. It would be unreasonable to expect it to describe all observable features of rainbows.

Supernumerary bows, as their name implies, are those in excess of the expected number. To understand them requires us to come to grips with the wave nature of light, to which geometrical optics is oblivious.

Alistair Fraser's article in the December 1983 issue of *Weatherwise* is the best exposition on supernumerary bows I have ever seen. It would be pointless for me to try to do better, so I have merely lifted one of his diagrams (Fig. 21.4), which shows the interference pattern produced when waves, rather than rays, are incident on a sphere.

In natural rainbows only a few supernumerary bows may be seen, three, perhaps four, at most. If you lust for more, they are easy enough to obtain in the laboratory. All you need is a hypodermic syringe or a hobbyist's oiler and a laser. If you use a hypodermic needle, you'll have to remove its sharp tip with a grinding wheel.

Fill the syringe with water and clamp it to a stand. With a bit of practice you'll be able to suspend drops from the tip of the needle. Transparent liquids other than water, such as syrups or oils, may be used.

In a darkened room illuminate a suspended drop, as large as possible, with

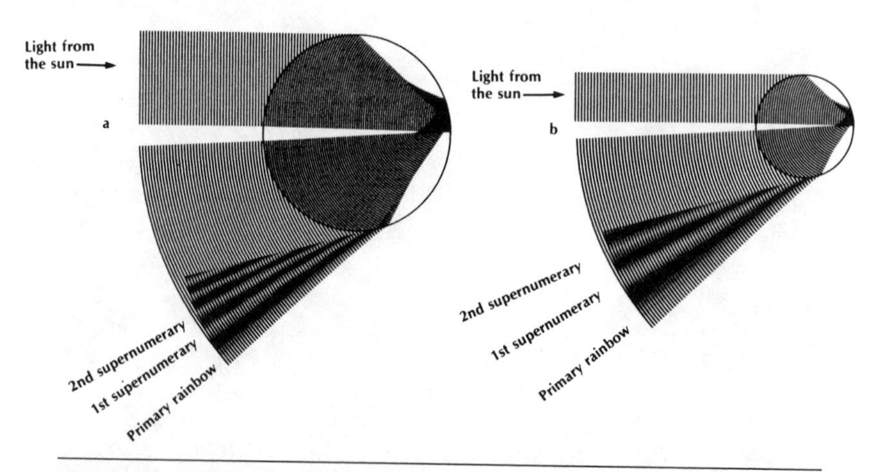

Figure 21.4 This diagram is the same as that in Figure 21.1 except that rays have been replaced with waves. Although rays cannot interfere, waves can, which is what gives rise to supernumerary bows. Figure courtesy of Alistair Fraser.

Figure 21.5 How many supernumerary bows do you count? These bows were obtained by illuminating a single water droplet with a laser beam. The broadest band is the primary rainbow, and it is accompanied by a series of supernumerary bows.

the laser beam. If you fiddle with the beam long enough—this takes patience because just as you get the beam aligned properly the drop falls off the needle and you have to squeeze out another—you may be rewarded with a beautiful set of supernumerary bows like that shown in Figure 21.5. These supernumerary bows, which have been projected onto a screen, lie inside the primary bow. Supernumerary bows also accompany higher-order bows. More fiddling with the beam may yield a set of supernumerary bows accompanying the secondary rainbow (not shown); these, of course, are not as bright as those accompanying the primary bow.

If the drop is not completely bathed in laser light only a partial bow will result. If you want a complete bow from a single drop, which may be obtained by expanding the beam, it will be at the expense of lower brightness.

As long as you have a drop suspended, you might as well explore another feature of rainbows, one that is not often observed (but not because it is rare). Both the primary and secondary rainbows are highly polarized (see Chapter 19 for a discussion of polarized light). You can verify this by inserting a polarizing filter in the laser beam and rotating it. And the next time you see a natural rainbow, look at it through polarizing sunglasses. By rotating them you can make it all but disappear.

WHY IS A RAINBOW CIRCULAR?

Why indeed? If it were not, this would be cause for much head scratching: How can more or less spherical raindrops illuminated by a nearly parallel beam of sunlight give anything other than a circular bow? Explaining the shape of the rainbow by appealing to the symmetry of a drop—it is spherical—is perfectly respectable. Yet rainbows in nature are *mosaics:* each part of a bow results from scattering by different raindrops. This is not at all like the demonstrations I have been discussing in which a bow, or portion thereof, is obtained with a *single* drop (or water-filled beaker). Rainbows in the laboratory are obtained by projecting the light scattered by a single drop onto a screen. Contrast this with natural rainbows, which result from scattering by a *myriad* of drops, each of which contributes its mite to what is observed.

Imagine looking at a sheet of rain illuminated by sunlight at your back. The raindrops scatter light in all directions but in some much more than in others. These special directions are determined by the rainbow angles. If we ignore the light scattered at angles other than the primary rainbow (my arguments apply equally well to the secondary rainbow), then each drop is the source of a cone of light the axis of which is parallel to the incident sunlight; the half-angle of this cone is the rainbow angle (strictly speaking, 180 degrees minus the rainbow angle). Each cone is composed of a bundle of rays, most of which are not directed toward your eyes. Those that are emanate from part of a circle. You can verify this with a paper cone and a long piece of string attached to its apex and fixed along its surface. A wall or a blackboard will serve to represent the sheet of rain. The loose end of the string may be held by a volunteer or tied to something. With its axis at a fixed angle to the wall, perpendicular, say, move the cone from point to point; all of these points will necessarily lie on a circle. This set of directions of the string represents those rays that intersect in a common point, the eye of the observer. Hence the circular shape of rainbows.

Once you realize that the rainbow is a mosaic, much more follows. For example, you are at the center of the rainbow you see, whereas someone nearby is at the center of the rainbow he sees. You and your companion therefore see different rainbows, that is, rainbows contributed by different sets of drops. Moreover, each part of your own personal rainbow originates from different drops: red from one set, green from another; the top of a bow from one set, the sides from another. If there are gaps in the sheet of raindrops there will be corresponding gaps in the rainbow. And since sunlight is a necessary ingredient in rainbows, drops in shade cannot contribute to them.

Complete rainbows (360 degrees) in rainshowers are possible, but to the best of my knowledge, one has never been photographed. Unless you are on a mountaintop or in an airplane looking down on a rainshower you will see at most 180 degrees of arc, and this only when the sun is on the horizon. Usually the earth prevents us from seeing that part of the rainbow below the astronomical horizon. When the sun is more than about 42 degrees above the horizon, the primary bow is not seen; when it is more than about 51 degrees, the secondary is not seen. You should be skeptical about tales of rainbows at noon—unless you encounter them in Icelandic sagas.

I have a suggestion for anyone who would like a bit of fame and fortune:

photograph a complete rainbow. You will need an airplane, better yet a helicopter because your view must be unobstructed, and a wide-angle lens on your camera. You will also have to persuade your pilot to fly in stormy weather. If you survive your flight you will have acquired something rare indeed.

22

Why Rainbows Are *Not* Impossible in Winter

both winters were spent in . . . repeating: "A noun is the name of a thing," which I had . . . heard my . . . teachers repeat until I had come to believe it.
U. S. Grant

Unequivocal statements about the supposed impossibility of wintertime rainbows can be found in at least two books. The author of one of them even went so far as to entitle an entire chapter "Why You Never See a Rainbow in Winter." Robert Greenler, the author of *Rainbows, Halos, and Glories,* knows better. On the dust jacket of his book there is a photograph of a rainbow arching over a snowcovered landscape in brazen defiance of proper rainbow behavior. Is Greenler's rainbow photograph a clever hoax? Or has he been the witness to an event of cosmic improbability?

Statements about the physical world which contain such words as "never" and "impossible" are to me compelling targets. When I see them I itch to refute them. Few events are truly impossible, although they may be rare. But I wouldn't classify wintertime rainbows—if I am allowed to stretch slightly what one means by a rainbow—as especially rare.

The incorrect, and easily refuted, notion that rainbows are impossible in subfreezing weather originates from a widespread misconception: (pure) water *necessarily* freezes at 0°C (32°F). We all have heard this so many times that we have come to believe it. Yet it would be more correct to say that 0°C is the melting point of ice than to say that it is the freezing point of water. But don't take my word for this, do an experiment to convince yourself that pure water can exist quite happily for long times at temperatures well below freezing.

A GOOD USE FOR TIN CANS

Several years ago I invited several students to my home for supper. During the course of the evening the conversation ranged over many topics. At some point I mentioned that a great many clouds are composed of liquid water droplets at temperatures well below freezing. I expected that heads would nod in agreement, but the reaction was shocked disbelief. So I fished an empty can out of the garbage, rubbed a thin coating of oil on its lid, and sprinkled several drops of water on it. Then I put the can in the freezing compartment of my refrigerator. At various intervals I removed the can. Although some of the droplets had frozen, many were obviously still in the liquid state.

Figure 22.1 Water in the form of small droplets can remain liquid for a long time at temperatures well below freezing. The drops shown in the top photograph were placed in the freezing compartment of a refrigerator (−11°C). After ten minutes, only about half of them had frozen (bottom), although they all were exposed to the same environment.

I subsequently have done this experiment with more care. For example, I have boiled the water to drive off dissolved air. The results of one such experiment are shown in Figure 22.1. Using an eyedropper I formed 17 water drops, all about the same size, on the lid of an orange juice can; a thin coating of oil on the lid causes the water to form as drops rather than as a film (see Chapter 7). Then I put the can in the freezing compartment of my refrigerator, the temperature of which was −11°C (12°F), for five minutes. Only six drops froze. After another five minutes two more had frozen. Then I put the can in the freezing compartment for ten minutes at the end of which time two drops stubbornly remained unfrozen.

I have done this experiment with drops of different sizes. The smaller the drops the longer they were likely to remain unfrozen at temperatures well below freezing. In the following sections I shall explain why this is so and what it has to do with rainbows.

SUBCOOLED WATER

The freezing of liquid water, like the condensation of gaseous water, is an example of a phase transition. In a gas, the molecules are in random, incessant motion, more or less independently of one another; each molecule is oblivious to what the others are doing. In a liquid, however, molecules spend part of their existence as members of clusters, which are continually forming and breaking up. Molecules in the liquid phase are not so independent as those in the gas phase. In a crystalline solid, molecular motions are even more severely constrained: the molecules form a regular array, the crystal lattice. In going from a gas to a liquid to a solid, a substance becomes more ordered. But it is not likely to do so spontaneously, it needs a nudge. Consider, for example, the condensation of water vapor into liquid droplets. In Chapter 2 I argued that, except at very high supersaturations, condensation requires the assistance of small particles called condensation nuclei. Merely increasing the relative humidity above 100 percent is not by itself sufficient to cause water vapor to condense. So it is also with the freezing of liquid water: some kind of foreign particle, an *ice nucleus,* is necessary to initiate freezing. This, like condensation of water on nuclei, is an example of *heterogenous* nucleation. And like condensation of water, freezing can also occur by *homogeneous* nucleation. If there are no ice nuclei in water its temperature can stay well below 0°C indefinitely. But if the temperature is lower than about −40°C (−40°F), the molecules become so sluggish that there is a high probability that enough of them will get together long enough to form a nucleus for further growth. At higher temperatures, say that of my freezing compartment, there is a finite but extraordinarily small probability that this will occur; it is not impossible, just highly unlikely, about as unlikely as all the air molecules in my room suddenly rushing into the corner, which is not impossible but, I am pleased to say, highly unlikely.

Water existing as liquid at temperatures below 0°C is often referred to as *supercooled,* although I prefer the term *subcooled.* This was advocated by the late James McDonald, who chose his words with great care. "Sub" means under or below, so it seems appropriate to refer to water at temperatures below 0°C as subcooled.

Before explaining the experiment, I should emphasize that I have implicitly assumed pure water drops. Dissolving a substance in water lowers its nominal freezing point, by which I mean the temperature of freezing provided that ice nuclei are present. This is why salt is sprinkled on icy sidewalks in winter and why a mixture of salt and ice is used to make ice cream at home. A simple explanation of why dissolved salt lowers the nominal freezing point of water is as follows. To freeze a liquid, the molecules must go from a partially ordered to a completely ordered state, which is accomplished by lowering the temperature sufficiently. But a substance dissolved in the liquid increases the disorder, hence the temperature must be lowered further to freeze the liquid. By way of analogy, suppose that you are given the task of arranging the members of a marching band, all of whom are wearing blue uniforms, in a regular formation. The band members are moving about randomly on a football field. Your task will be easier the slower they are moving. But now imagine that interlopers wearing white uniforms come onto the field. The task of formation is made more difficult, which can be compensated for by slowing everyone even further.

The existence of subcooled water does not depend on something being dissolved in it: subcooling occurs with very pure water. If you doubt this, you can do the experiments described previously with distilled water. And don't forget to boil the water to drive off most of the air dissolved in it. Like salt, or any dissolved substance, air does lower the freezing point slightly.

If, as I have stated, pure water can exist in the liquid phase more or less indefinitely at temperatures well below freezing, then why did some of the drops of my experiments freeze? The only explanation consistent with what I have said is that the drops must not have been free of nuclei. This should hardly come as a surprise. Tap water is really quite filthy stuff, potable to be sure, but teeming with all kinds of microscopic rubbish. And distilled water isn't really very clean. But even if one makes great efforts to prepare drops of really clean water, precautions must be taken to keep them clean because the surrounding air is filled with particles, some of which can serve as ice nuclei. And, of course, surfaces on which the drops are formed are not necessarily free of all nucleation sites. It is no wonder that some of my drops froze. I frequently opened and closed the door of the freezing compartment, and I did not take great pains to use very clean water and surfaces. From the previous paragraph, it should be fairly obvious why small drops are more likely to remain unfrozen. The larger the drop, the more likely it is to contain at least one ice nucleus (only one is necessary to initiate freezing of the entire drop). Even if formed from clean water, the larger drops are more likely to be struck by an ice nucleus in the surrounding air or to be formed on a nucleation site. A lake or pond or even a bucket of water is merely a huge drop in which there are almost certain to be some ice nuclei. This is why we have come to expect water to freeze at 0°C. Our experience is with the freezing of fairly large bodies of water, certainly much larger than cloud droplets and raindrops. It is the latter which give us rainbows, the former cloudbows, to which I now turn.

CLOUDBOWS

Rainbows are caused by particles that are more or less spherical. They must also be transparent. If spherical drops were to freeze, they would probably not be capable of causing rainbows, as evidenced by the cloudy appearance of the frozen drops shown in Figure 22.1. To convince yourself that a cloudy drop will not yield a rainbow, you can repeat the experiment described in the previous chapter using a drop of milk instead of clean water. Snowflakes, of course, do not cause rainbows. So the misconception that rainbows are impossible at subfreezing temperatures originates from the expectation that precipitation must fall either as snow or as frozen raindrops. But raindrops do not necessarily freeze at temperatures below 0°C, as those of us who suffer through ice storms every winter can testify. Freezing rain requires subcooled water drops. Such drops, when illuminated by sunlight, can yield rainbows. Although wintertime rainbows are rare, they are not impossible. Their rarity results in part from the nature of winter storms: no matter how intense the rainfall, you can't see a rainbow if the sky is covered with clouds. If rainbows are more frequent in warmer months, it is partly because this is when local convective storms occur: they provide rain but often don't completely obscure the sun.

For the sake of argument, let us grant that rainbows are rare unless temperatures are above freezing. There is, however, a variation on the rainbow, the cloudbow or fogbow (see Fig. 22.2), which as often as not is observed when temperatures are *below* freezing. All bows are not equal. The purity of their colors depend very much on the sizes of the drops that cause them, although this is not evident from the elementary theory of the rainbow (see the previous chapter). Thus a family of bows exists; some members display striking colors, others are more pastel. As a rule, the larger the droplets the purer the colors. So bows formed by cloud droplets are not very colorful; their beauty is more subtle than that of their gaudier cousin the rainbow.

Cloud droplets are so small that they can remain unfrozen for long times even at temperatures well below freezing. So it is no trick at all for a cloudbow to be caused by subcooled water droplets. Indeed, a great many clouds, especially at middle and high latitudes, are composed of such droplets. And if you see a cloudbow, you are likely to be in an airplane. So I think that it is not an exaggeration to say that cloudbows are observed often when temperatures are below freezing.

Now I can discuss something that I wouldn't have dared to earlier. Figure 1.2 is a photograph of a cloud formed in the neck of a freshly-opened bottle of beer. I asserted that it formed by homogeneous nucleation, which I supported

Figure 22.2 Both rainbows and the cloudbow, the angular positions of which are about the same, are caused by water droplets illuminated by sunlight. Because cloud droplets are much smaller than rain drops, cloudbows display less vivid colors than rainbows. For the same reason, supernumerary bows associated with cloudbows have a greater angular width than those associated with rainbows. Photograph courtesy of Francis Turner.

by adducing my calculations of the extreme low temperature to which the gas in the neck dropped momentarily when the bottle was opened. This probably led you to conclude—and who would have blamed you?—that this cloud was composed of ice particles. Yet I doubt that this was so, and for two reasons. First, I pointed out in Chapter 3 that condensation is a mechanism for heating. The cloud particles in the bottle formed by condensation, hence were warmer than their surroundings, although not above the freezing temperature. But, as I have tried to convince you, tiny droplets of water can exist quite happily in the liquid state at temperatures well below what we call the freezing point.

I hope that as a consequence of this chapter I will be deluged in the years to come with photographs of rainbows in the depths of winter. There is nothing like a healthy dose of reality to demolish notions of what is or is not possible.

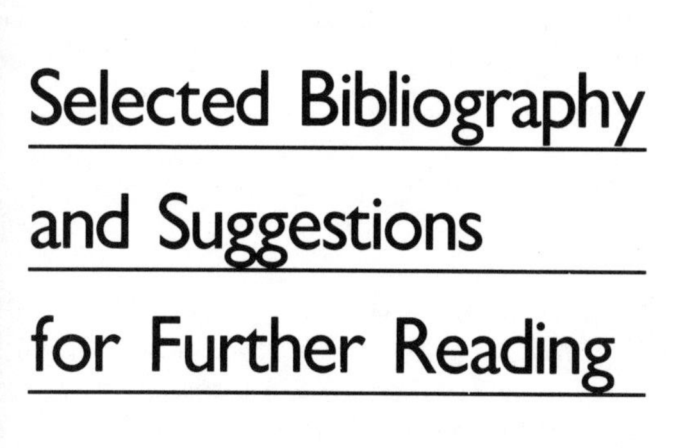

Selected Bibliography and Suggestions for Further Reading

I wove some references into the text, especially if I made direct use of them or they greatly influenced my thinking. If I have whet your appetite, you might find more fare to your liking in some of the following books and papers. I do not include in this list any references cited in the text. These are just ones that didn't seem to fit at the time I was writing the original articles or that appeared subsequently or that I became aware of only subsequently.

GENERAL: My book is a collection of essays in which I discuss in depth the physics of some readily observable phenomena. Jearl Walker's celebrated *Flying Circus of Physics* (Wiley, 1977) is a collection of questions about many such phenomena, brief answers to them, and a long list of references which readers can go to for more details. My book therefore complements his. I discovered only recently a delightful Russian book (by way of a Spanish translation), *Physics for Entertainment,* by Yakov Perelman, which is similar in some ways to both this book and to Walker's. It also has been translated into English (Mir Publishers, 1972).

CHAPTER 1: There are a surprising number of people interested in bubbles, including oceanographers, chemical engineers, and physiologists. An entire book could be devoted to bubbles; indeed, part of an advanced treatise has been: *Bubbles, Drops, and Particles* by R. Clift, J. R. Grace, and M. E. Weber (Academic, 1978).

CHAPTER 2: Louis Battan's *Cloud Physics and Cloud Seeding* (Doubleday, 1962) is a good elementary book on cloud physics. A much more advanced and comprehensive treatise on all that was known about the physics of clouds before 1980 is *Microphysics of Clouds and Precipitation* (Reidel) by H. R. Pruppacher and J. D. Klett.

CHAPTER 3: A mostly elementary treatment of humidity, including its measurement, is H. L. Penman's *Humidity* (Institute of Physics, 1955).

CHAPTER 4: Raoult's paper and other classics of solution chemistry were translated by Harry C. Jones in *The Modern Theory of Solutions* (Harper, 1899).

I am not alone in wincing every time I see or hear about warm air holding more water vapor than cold air. See the letter by Robert E. Lautzenheiser in the June 1986 issue (p. 133) of *Weatherwise*.

The molecular interpretation of vapor pressure reduction is clearly stated by J. H. Hildebrand and R. L. Scott on pages 17 and 18 of *The Solubility of Nonelectrolytes*, 3rd ed. (Reinhold, 1950).

CHAPTER 6: The discovery of atmospheric pressure and the invention of the barometer has been discussed admirably by W. E. Knowles Middleton in his treatise *The History of the Barometer* (Johns Hopkins, 1964).

H. G. Cowling's *Molecules in Motion* (Harper, 1960) is an introduction to the kinetic theory of gases by one of the pioneers in the field. It requires no mathematical knowledge beyond algebra. A somewhat more advanced treatise, although still without all the heavy mathematical artillery, is Sir James Jeans's *An Introduction to the Kinetic Theory of Gases* (Cambridge U. P., 1940).

CHAPTER 7: In writing this chapter I profited greatly from the following papers, listed in order of increasing level of treatment: M. V. Berry, "The molecular mechanism of surface tension," *Physics Education*, Vol. 6, 1971, p. 79; R. C. Brown, "The surface tension of liquids," *Contemporary Physics*, Vol. 15, 1974, p. 301; R. C. Brown, "The fundamental concepts concerning surface tension and capillarity," *Proceedings of the Physical Society*, Vol. 59, 1947, p. 429.

Detailed instructions for building a dewpoint hygrometer are given by Hampton W. Shirer in the June 1986 issue (p. 160) of *Weatherwise*.

Sean Twomey's results were reported in "Experimental test of the Volmer theory of heterogeneous nucleation," *Journal of Chemical Physics*, Vol. 30, 1959, p. 941.

CHAPTER 10: In *Electricity and Matter* (Scribners, 1904), J. J. Thomson, the discoverer of the electron, gave a simple discussion of radiation by an accelerating charge. A more thorough discussion along the same lines was given many years later by J. R. Tessman and J. T. Finnell, "Electric field of an accelerating charge," *American Journal of Physics*, Vol. 35, 1967, p. 523.

CHAPTER 12: What I know about colorimetry I have learned mostly from the following books: *The Science of Color*, published by the Optical Society of America in 1963; *Principles of Color Technology,* 2nd ed. (Wiley-Interscience, 1981) by F. W. Billmeyer and M. Saltzman; and *Color Measurement,* 2nd ed. (Springer, 1985) by D. L. MacAdam.

Purely by chance I recently happened upon "Once in a blue moon" by R. L. Reese and G. T. Chang in the March 1987 issue of *Griffith Observer.* They discuss both the physics of real blue moons and the blue moons of astronomers—which aren't blue.

CHAPTER 13: A good concise discussion of the green flash was given by Alistair Fraser in the August 1980 issue (p. 173) of *Weatherwise.* An entire book, D. J. K. O'Connell's *The Green Flash and Other Low Sun Phenomena* (North Holland, 1958), is devoted to this topic; it is valuable mostly for its color plates and its history of investigations of the green flash. The most thorough analysis of the green flash was done by Glenn Shaw: "Observations and theoretical reconstruction of the green flash," *Pure and Applied Geophysics,* Vol. 102, 1973, p. 223. Alistair Fraser discusses the atmospheric temperature profiles necessary for a green flash in "The green flash and clear air turbulence," *Atmosphere,* Vol. 13, 1975, p. 1.

CHAPTERS 14 and 15: I discuss multiple scattering at an intermediate level in "Multiple scattering of light and some of its observable consequences," *American Journal of Physics,* Vol. 55, 1987, p. 524.

CHAPTER 16: The classic treatise on visibility in the atmosphere is W. E. K. Middleton's *Vision Through the Atmosphere* (University of Toronto, 1952).

CHAPTER 17: For further discussions of coronas and iridescence (and many other atmospheric optical phenomena) I highly recommend *The Nature of Light and Color in the Open Air* (Dover, 1954) by M. Minnaert and *Introduction to Meteorological Optics* by R. A. R. Tricker (Elsevier, 1970).

CHAPTER 19: At long last a popular book on polarized light in nature, with many color plates, has been published: Gunther Können's *Polarized Light in Nature* (Cambridge U. P., 1985).

Anyone interested in the history of polarized light must read Edwin Land's paper "Some aspects of the development of sheet polarizers," *Journal of the Optical Society of America,* Vol. 41, 1951, p. 957. A history of optics, including polarization, up to the beginning of this century is Ernst Mach's famous book *The Principles of Physical Optics* (Dover).

CHAPTER 21: The rainbow in art is discussed in Raymond Lee's articles "Rainbows: An observer's guide," "God, the rainbow, and the artist," "What's at the end of the rainbow puzzle?" published simultaneously in *The Johns Hopkins Magazine, Franklin and Marshall Today, The Wick, At Renssaelaer,* and *The WPI Journal,* August 1984, p. i.

CHAPTER 22: Experimental evidence that liquid droplets rather than ice particles usually form in expansion chambers is given in B. M. Cwilong's paper "Sublimation in a Wilson chamber," *Proceedings of the Royal Society*, Vol. A190, 1947, p. 137.

Another simple demonstration of subcooling is discussed by Ira Geer in the April 1979 issue (p. 84) of *Weatherwise*.

Index

S. GUILDFORD,
C. ALLISON